Synesis

The complexity of today's large organisations, businesses, and social institutions defeats management approaches based on monolithic thinking. Most industry and service organisations look at their performance either from a single perspective – productivity, quality, safety, etc. – or from different but separate perspectives that reside in organisational silos. Quality is treated separately from safety, which, again, is treated separately from productivity, and so on. While siloed thinking may be convenient in the short term, it fails to recognise that any specific perspective reveals only a part of what goes on. Yet it is essential to have a unified view of how an organisation functions effectively to manage changes and to ensure the organisation excels in what it does.

Synesis represents the mutually dependent set of priorities, perspectives, and practices that an organisation needs to carry out its activities as intended. It shows how to overcome the fragmentation in foci, scope, and time that characterises the dominant change management paradigms. This book is consequently not about productivity or quality or safety or reliability but about all of these together. It is about why it is necessary to think of them as a whole. And it is about how this can be done in practice.

Erik Hollnagel is Senior Professor of Patient Safety at Jönköping University (Sweden) and Visiting Professorial Fellow at Macquarie University (Australia). He has throughout his career worked at universities, research centres, and with industries in many countries and with problems from a variety of domains and industries. Erik is an internationally recognised expert in the fields of safety and complex systems analysis and has helped create the fields of resilience engineering, resilient health care, and Safety-II. He has published widely and is the author/editor of 27 books as well as 500+ papers and book chapters.

Synesis

The Unification of Productivity, Quality,
Safety and Reliability

Erik Hollnagel

Routledge
Taylor & Francis Group

LONDON AND NEW YORK

First published 2021
by Routledge
2 Park Square, Milton Park, Abingdon, Oxon OX14 4RN

and by Routledge
52 Vanderbilt Avenue, New York, NY 10017

Routledge is an imprint of the Taylor & Francis Group, an informa business

British Library Cataloguing-in-Publication Data
A catalogue record for this book is available from the British Library

Library of Congress Cataloging-in-Publication Data
Names: Hollnagel, Erik, 1941– author.
Title: Synesis : the unification of productivity, quality, safety and reliability /
    Erik Hollnagel.
Description: Abingdon, Oxon ; New York, NY : Routledge, 2020. | Includes
    bibliographical references and index.
Identifiers: LCCN 2020012905 (print) | LCCN 2020012906 (ebook) |
    ISBN 9780367481490 (hardback) | ISBN 9781003038245 (ebook)
Subjects: LCSH: Organizational effectiveness. | System analysis. | Management.
Classification: LCC HD58.9 .H6535 2020 (print) | LCC HD58.9 (ebook) |
    DDC 658.5—dc23
LC record available at https://lccn.loc.gov/2020012905
LC ebook record available at https://lccn.loc.gov/2020012906

ISBN: 978-0-367-48149-0 (hbk)
ISBN: 978-1-003-03824-5 (ebk)

Typeset in Bembo
by Apex CoVantage, LLC

Cover artwork by Erik Hollnagel & Yushi Fujita (calligraphy). The meaning of the
kanji, pronounced soku or tabaneru in Japanese, shù in Chinese, is a collection of
things that are bundled.

Contents

vi   *Contents*

1 A fragmented view

Introduction

There are many ways to characterise the organisations and companies that constitute the fabric of present-day society, no matter what their purpose is and regardless of whether they are in industrialised or developing countries. Organisations have to be dynamic, creative, efficient, competitive, fast-paced, agile, and resilient – and possibly other things as well. One consequence of this is that an organisation today no longer can be left to itself to develop and mature at a natural pace in relative independence from the surroundings in which it exists. Internal and external changes, as planned improvements or as unplanned interruptions and disruptions, are today so frequent and develop so fast that a natural pace of development simply is inadequate. To thrive and survive it is essential that an organisation's ability to change matches the nature of the events that may force or require a change.

A widespread practice for thinking of and managing organisations is to make a distinction between the internal factors that an organisation has some control over and the external factors that typically are things an organisation cannot control (cf. the discussion of boundaries in Chapter 7). The internal factors include tangibles such as the technology that provides the infrastructure and the means by which organisations work and produce, as well as intangibles such as staff morale, company policy, and culture. The external factors are many and varied and include resources of one kind or another that an organisation needs, rules and regulations imposed by others (e.g., by legislation), and unexpected events such as natural disasters or economic, technological, or political disruptions. The external factors also include social factors and social trends (e.g., pressure groups or changes in preference or needs), competition (relative or absolute), economics on both macro and micro levels, and, again, technology. The internal factors should preferably be under control, while the external factors usually are not. Management can therefore be seen as an attempt to cope with the internal and external unpredictability where the former in principle can be controlled while the latter cannot.

The management of how a system or organisation (the two terms will be used interchangeably throughout the book) performs is always relative to one or more issues or criteria. A criterion is in this context understood as a reference that describes what should happen both in terms of the specific characteristic and in terms of its desired magnitude. A criterion should not be confused with a mission statement. The latter describes why an organisation exists; what the goal of its operations is, such as the kinds of products or services it provides; and who the primary customers, clients, or market are. It often includes a short description of an organisation's values or philosophies and the desired future state – an organisation's "vision." Where the mission statement describes the *raison d'être* of an organisation, the criterion describes the reference for judging how well it works. The missions statement for an airline could be "X is a global company dedicated to providing air transportation services of the highest quality and to maximising returns for the benefit of its shareholders and employees"; but when you are on board, the captain will most likely greet you by emphasising that "safety is our highest priority." The business may be to transport people, but the criterion – or one of the several criteria – is to do it safely. (At one point during Boeing's struggle to overcome the consequences of the problems with the 737 MAX, the former Boeing chairman and chief executive Dennis Muilenburg declared, "Nothing is more important to us than the safety of the flight crews and passengers who fly on our airplanes." This was in July 2019, when Boeing took a \$4.9 billion hit to cover costs related to the worldwide grounding of its 737 MAX aircraft. In December 2019 Boeing announced that it would temporarily halt production of its troubled 737 MAX airliner beginning January 2020.)

Safety is an important issue for many organisations, whether they produce something or provide services, but, of course, not the only one. Quality is another important issue, not least for companies that depend on customers to purchase their products or services. Productivity – or efficiency – is a third issue, although it may be more important for an organisation and its investors than for its customers. And so on . . .

It is common knowledge, of course, that no issue can be considered in isolation from the others. Organisations that nevertheless do so are categorised as pathological or calculative. A pathological organisation focuses on a single issue – for instance profitability – and does not even bother to consider any others. A calculative organisation has decided that the benefits of pursuing a single issue outweigh – and even justify – the cost of neglecting any others. Neither of these is considered a good organisation. Yet even organisations that try to take multiple issues into account tend to do it one by one rather than considering them together. This can easily be seen from the organigram of any company large enough to have one, which typically will look something like the generic organigram shown in Figure 1.1.

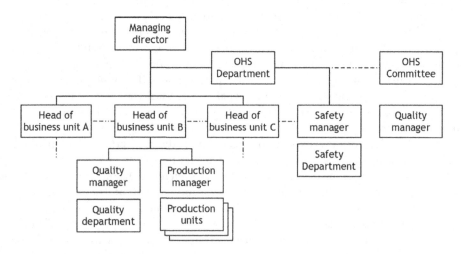

Figure 1.1 A typical organigram.

In the case of the (fictive) organisation shown in Figure 1.1, each business unit has a production manager and a quality manager. These report through the line to the head of their unit, who in turn reports to the managing director. There is also a safety manager who presumably is responsible for safety across the business units as well as for the organisation as a whole. The safety manager reports to the OHS department and liaises with the OHS committee. What is typical for this way of organising a company is that there are a number of specialised management positions in the organisation. Each has its own role, people, resources, and competence, and each has a specific focus and performance criteria. Each may report to a common point or role higher up in the organisation – in this case the managing director – but each position has a different focus and works independently of the others rather than together.

- For some organisations safety comes first, although it is never safety regardless of productivity. Safety, at least from a Safety-I perspective, is protective rather than productive, hence a cost rather than a source of revenue. So if nothing is produced, there will be no resources for safety.
- For some organisations productivity comes first, although that cannot be regardless of safety or quality. One consequence of inadequate safety is the possibility of accidents and incidents that will weaken or perhaps even disrupt production. Hence, if there was no safety, productivity would be jeopardised. Similarly, insufficient quality will reduce market share and therefore also affect productivity.
- For some organisations quality comes first, but, again, it cannot be quality regardless of productivity or, for that matter, regardless of safety. Quality assurance may improve profitability but does not as such increase productivity. By contrast, inferior quality will perturb the production flow and therefore have much the same effect as a lack of safety.
- For some organisations reliability comes first, but as in the other cases it cannot be reliability regardless of productivity. Indeed, productivity – as well as safety and quality – may depend heavily on the reliability of the technology as well as the people who provide the basis for what an organisation does.

The same kind of argument can be made for how safety depends on quality, how safety depends on reliability, how productivity depends on quality, how productivity depends on reliability, and so on. Altogether it is unwise – and in the long run also impossible – to manage an organisation without considering how the various issues relate to each other in either a synergistic or antagonistic way. It is nevertheless typical for how organisations are managed. The remedy is simple: consider the issues together rather than one by one. The problem is how this can be done in a practical way. That is precisely what this book is about.

Productivity, quality, safety, and reliability

At this point the reader may well wonder why the discussion refers to these four issues rather than four others, or why it is four issues rather than five, or six, or more. The reason is simply that the four issues represent something that is essential for practically all organisations. The issues may differ in weight or importance, depending on the type of business or service that is the main focus of an organisation. For a hospital, for instance, quality and productivity may be more important than safety and reliability, concerns about patient safety notwithstanding. For an airline, safety usually comes first – at least in relation to the travelling public. For an engineering company or a public service provider – for instance in telecommunication – reliability of the services may be rated as most important. And for a company manufacturing consumer goods, quality may seem to be more important. The relative importance may furthermore depend on the time horizon of an organisation and its management, where the short term often seems to be given more attention than the long term. The reason is also pragmatic. The four issues are clearly not the only ones. Other issues could be sustainability, customer or user satisfaction, the well-being of the workforce, environmental impact, etc. Extending the list does not, however, change the fundamental argument – that it is necessary to consider them together rather than separately; it only makes it a bit more cumbersome. Having said that, the reader is welcome throughout the book to replace one issue with another, or to add further issues.

Examples of what can happen when one issue dominates and therefore excludes other issues are easy to find, such as the City Circle Line described below.

The City Circle Line

The City Circle Line is a loop of the Copenhagen Metro, claimed to be the largest construction project to have taken place in Copenhagen, Denmark, during the last 400 years. The building of this extension to the already existing metro lines was approved by the Danish Parliament in 2007, and construction started in 2013. The opening of the City Circle Line was originally planned to happen in December 2018 but was postponed until July 2019. It actually took place in September 2019 (needless to say, there was a budget overrun as well).

When the construction began, the target of the Copenhagen Metro Team (CMT) was a maximum of 16 accidents per 1 million hours worked. In July 2019 it was revealed that the actual number of accidents was about 25% higher – namely, 20.6 accidents per 1 million hours worked. The reason is not hard to find. Subcontractors were required to meet strict deadlines to reduce the delays. The leading issue was therefore productivity rather than safety. One worker who had his left forearm crushed by a large block of concrete explained that people were under considerable pressure to work as fast as possible. Asked whether he could have refused, the answer was that if he had done so the site managers would have just found someone else to take his place and he, as a result, would have been without a job.

The origins of the fragmented view

Experience strongly suggests that it is a bad idea to let one issue take precedence over the others – even if the organisation does not regress to a calculative state. Most organisations have realised this and therefore try to pursue several issues at the same time, although in a fragmented way, as illustrated by Figure 1.1. An organisation may well have a safety department, a quality department, several production units, etc. But the departments function independently in strenuously protected functional silos. This makes it difficult for employees or even departments to share information or knowledge with each other and often leads to a way of working where information sharing is the exception to the rule. While this may help keep staff focused on their areas of expertise and reduce distractions, it can also lead to a plethora of internal and external problems for everyone involved, including a resistance to change, difficulties in communication and collaboration across units, repetition of work or conflicting goals, unnecessary redundancy, and poor decision-making. Despite this it seems to be the rule rather than the exception that different issues are pursued separately, or, in other words, that organisations have a fragmented rather than unified view. But why is this so? There are at least two main reasons for this, one historical and the other psychological.

Historical reasons for fragmentation

It is practically a universal characteristic of humans that we deal with problems when and as they happen rather than in advance. This is also the case for the four issues which, from a historical perspective, achieved prominence at different times (cf. Figure 1.2). In the figure the appearance of each issue is linked to the publication of either a report or a book. That does not mean, of course, that they appeared *deus ex machina* at that specific moment in time, nor that they jumped out in final form like Athena born fully armed from the forehead of Zeus. The problems had in each case been known for a long time, but for each it was also the first time that a solution had partly or wholly been able to solve the problem – at least for a time – and thereby allow appreciable and possibly significant progress to be made. In each case it was also the first time a coherent theoretical treatment had become generally accepted by practitioners looking for a solution. So, for the sake of argument, it seems reasonable to mark the point in time when each concept gained public recognition as the moment of birth.

Productivity was the first issue to attract attention and came to the fore in the guise of the scientific management movement. Scientific management is an approach to work that analyses and synthesises workflows, with the main objective of improving economic efficiency and labour productivity. It is often referred to as Taylorism after its founder, Frederick Winslow Taylor, and we may therefore conveniently date it to the publication of Taylor's monograph *The Principles of Scientific Management* in 1911.

Quality and safety next came onto the scene, and coincidentally in the same year. The concern for quality is marked by the publication of Walter A. Shewhart's book *Economic Control of Quality of Manufactured Product* in 1931. The declared aim of that work was to develop the scientific basis for economic control of the quality of manufactured products. (Interest in the quality of a product has obviously always been a concern and therefore goes as far back in time as humans have created artefacts. The same can, with equal justification, be said of productivity and safety.) Shewhart was a statistician working in the

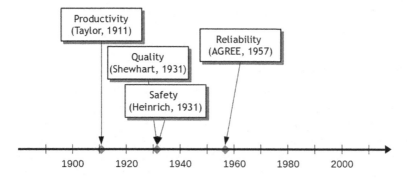

Figure 1.2 Historical reasons for the fragmented view.

Inspection Engineering Department at Bell Telephone Laboratories. The systematic concern for safety is similarly marked by the publication of a book entitled *Industrial Accident Prevention: A Scientific Approach* by Herbert William Heinrich in 1931. Heinrich was assistant superintendent of the Engineering and Inspection Division of Travelers Insurance Company and therefore had an obvious interest in understanding why accidents in the industry happened and how they could be prevented.

In hindsight it would seem reasonable or even natural if the two books and the two authors were related and aware of each other. Safety and quality certainly go together even if they usually belong to different departments or silos. There is, however, nothing to suggest that Heinrich and Shewhart knew each other or even that they knew about each other. This is perhaps yet another symptom of a fragmented view.

The fourth criterion – reliability – is marked by a report called *Reliability of Military Electronic Equipment* , published by the Advisory Group on Reliability of Electronic Equipment (AGREE) in 1957. AGREE was jointly established in 1950 by the United States Department of Defense and the American electronics industry. The group "provided all the armed services with the assurance that reliability could be specified, allocated, and demonstrated; i.e. that a reliability engineering discipline existed" (Saleh & Marais, 2006). Reliability has, of course, always been a concern as a human quality, typically associated with the presence of trust. But AGREE was concerned about reliability as a technical rather than human quality. The need came from the rapid introduction of electronic components and devices in systems – first military and later civilian. Although components in the beginning were not very reliable, people and organisations – not least the military – soon became totally dependent on them and hence needed to be reassured of their reliability. Each of the four issues will be described in more detail in Chapter 2.

With the aid of hindsight it seems that when productivity, or efficiency, became the focus of *Scientific Management* in 1911, quality and safety were problems on a smaller scale. Accidents have obviously always happened – the origin of the word *accident* is the Latin verb *accidere*, meaning "to fall down" or "to fall" – but one hundred years ago industrial accidents were neither so frequent nor so serious that they were seen as unacceptable or unaffordable and hence were tolerated in a way that may be hard to understand today. Neither was quality yet an issue of the same magnitude as productivity. Safety and quality both became concerns a couple of decades later, not least because the productivity problem had been solved – or at least brought sufficiently under control so that it no longer was a major issue. And, finally, reliability came onto the scene not once but thrice. In all cases the issues became conspicuous because they were seen as a business problem, which basically means they stood in the way of greater profit. The motivation in each case came from cost and efficiency concerns rather than a concern for, e.g., the well-being of people or the greater good of society.

Psychological reasons for fragmentation

Another reason for the fragmented view is psychological and has to do with the tradition of understanding phenomena through decomposition, i.e., by taking a complex item and breaking it into its elements. The idea that matter is made up of discrete units appears in many ancient cultures such as Greece and India. The word *atomos*, meaning "uncuttable," was coined by the ancient Greek philosophers Leucippus and his pupil Democritus (c. 460–370 bc). Democritus taught that atoms were infinite in number, uncreated, and eternal and that the qualities of an object result from the kind of atoms that compose it.

Decomposition is indeed a fundamental feature of the physical world as the process by which organic substances are broken down into simpler organic matter.

Decomposition is also the way in which we like to explain something, i.e., by referring to the parts and then to the parts of those parts and so on until we reach the indivisible atom – or perhaps the Higgs boson. In Western scientific thinking the general principle is to divide problems into smaller, constituent problems and then reason about them, and to continue to divide problems into smaller problems until they are of a magnitude that is thought to be solvable.

One consequence of decomposition is a continuously growing number of details and increasing specialisation. This will in practice soon lead to departmentalisation, where each department, or silo, becomes better and better at its speciality but also increasingly isolated from other specialities. The attractive consequence, from a managerial point of view, is that departmentalisation and self-containment reduce the need for coordination and hence make multiple activities easier to manage. In this way the psychological pre-dilection for decomposition, which possibly is due to our cognitive make-up, in that it is simpler to keep track (mentally) of a few things than of many, carries over into manage-ment, which, of course, also can be seen as a cognitive activity. We thus end up trying to solve every problem we meet with decomposition, including the problems of managing an organisation and its performance. The psychological reasons for the fragmented view will be described in further detail in Chapter 3.

Tractable, intractable, and entangled systems

Decomposition has always been an attractive approach, as proven by the classic divide and rule (Latin: *divide et impera*), or divide and conquer, principle. Divide and rule has been used in politics, warfare, and business to gain and maintain power. It works because the principle of breaking larger concentrations of power into smaller parts that individually have less power makes them easier to control or manage. Looking at something in terms of its parts or fragments rather than as an aggregated whole also reduces the cognitive workload; instead of having to consider many things at the same time, they can be looked at one by one.

Fragmentation or decomposition – dealing with something part by part and step by step rather than together – has historically been so successful that it almost seems to be the only way to manage something. It worked effectively until about a century ago because systems and organisations until then were tractable in the sense that it was possible to follow what happened and understand how they worked. Because of that they were also (relatively) easy to manage. For a system to be manageable there must be a reasonably clear description or specification of what it is and how it works – of what goes on "inside" it. This is also necessary to analyse a system, whether for risks or in *post hoc* event analyses (accident investigations). The reason is simply that it is impossible to manage something effectively unless it is clearly described and unless the modes of functioning are known. Neither is it possible to imagine what can happen – risk assessment – nor to produce factual rather than fictitious explanations of why something has happened without that knowledge. Table 1.1 summarises the characteristics of tractable and intractable systems.

A system is tractable if it is simple to describe, if most of what happens is observable rather than hidden, if the principles of functioning are known, and if it does not change before a description can be produced. Conversely, a system is intractable if it is difficult to describe, either because there are many details or because much of what happens cannot

Table 1.1 Characteristics of tractable and intractable systems

	Tractable system	*Intractable system*
Number of parts and details	Few parts and simple descriptions with limited details.	Many parts and elaborate descriptions with many details.
Comprehensibility	Principles of functioning are known. Structure (architecture) is also known.	Principles of functioning are known for some parts but not for all.
Stability	Changes are infrequent and usually small. The system is in practice stable and can therefore be fully described.	Changes are frequent and can be large. The system changes before description is completed.
Dependencies among system parts and functions	System parts and functions are relatively independent and only loosely coupled.	System parts and functions are mutually dependent and tightly coupled.
Relationship to other systems	Independence, low level of vertical and horizontal integration.	Interdependence, high level of vertical and horizontal integration.

be observed; if the principles of its functioning are only partly known or even completely unknown; and if it changes before a description can be completed.

The foundations for the management practices we use today are from a time when most systems and organisations were tractable. Unfortunately, that is no longer the case, and it has consequently become increasingly difficult to manage many of the systems that we unintentionally have made ourselves dependent on. It is therefore worthwhile to consider for a moment why and how this transition from tractable to intractable took place.

The answer is actually fairly simple. It happened because systems and organisations have never been perfectly or fully specified. Because of that something will happen sooner or later for which there are no prepared responses, often because such events were impossible to imagine but sometimes because the time or resources allocated to carry out a thorough analysis were insufficient. Ron Westrum has elegantly described this as a problem of requisite imagination:

> When we reflect on how seemingly minor flaws can have such disastrous outcomes, the importance of anticipating what might go wrong becomes apparent. The fine art of anticipating what might go wrong means taking sufficient time to reflect on the design in order to identify and acknowledge potential problems. We call this fine art requisite imagination.
>
> (Adamski & Westrum, 2003)

Even earlier, Merton (1936) had formulated his elegant Law of Unanticipated Consequences, described further in Chapter 4, which states that there will always be consequences or outcomes that have neither been foreseen nor are intended by purposeful action.

When something unexpected happens, the natural response is to try to understand why so that steps can be taken to prevent it from happening again. This "find-and-fix" attitude makes efficiency, and particularly the speed of response, more important than thoroughness. Solutions are therefore often symptomatic patches or Band-Aids that "solve" the most pressing problems but that rarely go into depth. Such solutions not only are

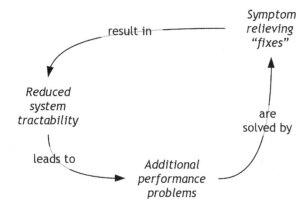

Figure 1.3 The problems with "fix-and-find."

incomplete but also reduce the tractability of the system. This quickly creates a vicious circle that gradually makes the system more intractable (cf. also the discussion in relation to Figure 4.3 later): "[h]ope drives us to invent new fixes for old messes, which in turn create even more dangerous messes" (Wright, 2004, p. 123).

We thus routinely try to solve problems one by one without admitting or even being concerned about the fact that they are not solvable in that way. We generally fail to understand how functions are interconnected and that attempts to manage them without taking the interconnections into account will only make matters worse. As the number of systems and functions continue to grow and as both become ever more interdependent, systems will sooner or later move beyond intractability and become, for lack of a better term, entangled. This term is borrowed from quantum theory, where it is defined as the physical phenomenon that occurs when pairs or groups of particles are generated, interact, or share spatial proximity in ways such that the quantum state of each particle cannot be described independently of the state of the others, even when the particles are separated by a large distance. When applied to conventional systems and organisations this means that their functions, as well as the systems themselves, have become intrinsically linked or intertwined. Not only do systems have an exceeding number of parts and require elaborate descriptions with many details, but no part can be described or understood unless all the other parts are also accounted for. The fact that there are many parts with many interconnections also means that more things will happen at the same time.

One characteristic of quantum entanglement is that the effects can occur instantly. While that is not the case in the physical world, it is often surprising how fast something can happen – for instance how fast news and rumours travel. Things can happen very fast both because communication is faster and because there are unrecognised channels or couplings in organisations – the so-called small world problem (Milgram, 1967) that seems to characterise social networks. Paradoxically, something may also happen so infrequently or change so slowly that it is not recognised. Currently, the prime example of this is global warming.

It is essential for any organisation that it is able to manage the conditions necessary to sustain its existence. In other words, it must be able to manage the surroundings and the problem it faces. It must be able to cope with the challenges and opportunities of the world in which it exists. Yet humans are way past the point where they can successfully cope with the complexity of the world they themselves have created. Humans have a limited cognitive capacity that cannot be expected miraculously to grow or change so that everything suddenly becomes manageable again. The point has also been passed where problems can be solved by making the world simpler or so simple that it becomes comprehensible – partly because we as humans are part of that world. Pinning our hopes on some kind of new technology, such as AI, to solve the problem for us is futile since we, to begin with, would be unable to understand the technology or AI itself (as copious examples show).

It is nevertheless necessary to find a way to describe the world that is both comprehensible and able to make the challenges manageable; more will be said about that in Chapter 4. It is necessary because we, in the end, are the ones who have to cope with the problems. The approach advocated here is to think in terms of functions rather than structures, to think in terms of processes rather than outcomes (or products), and to think in terms of change and dynamics rather than stability and statics. Chapters 2–6 that follow will introduce the details of this approach, called synesis, while Chapter 7 will try to bring it all together in a nexus of necessary knowledge.

2 Historical reasons for the fragmented view

Introduction

As described in Chapter 1, the four issues considered here – productivity, quality, safety, and reliability – came to the fore at different points in times. It is beyond the scope of this book to look into how and why this happened, even though that might provide a fascinating account of how Western industrialised societies developed. Figure 1.2 shows the chronology in terms of the four publications that mark the introduction of each issue to both public awareness and the corporate worlds. All four had obviously been a concern for people and societies much earlier, although not necessarily by the same names or labels.

One way of explaining how something becomes an issue is to look at it in terms of the signal-to-noise ratio (S/N). The signal represents the desired and intended outcome of what is being done – for instance in a factory or by a service provider. If the signal cannot be clearly distinguished from the noise – in other words, if the results are not clearly recognisable, if there is too much background noise or too much unwanted variability – the obvious solution is to reduce or eliminate the noise. Productivity was, and still is, an issue because it was, and still is, necessary for a company to be economically viable, especially if there are large investments at play. If, therefore, productivity was unsatisfactory, the solution was to look for the noise that "drowned" the productivity signal, so to speak. Once this had been addressed and the signal had become clearly distinguishable, other sources of noise became noticeable, primarily inadequate quality and safety. When the "noise" from both quality and safety had been reduced to an acceptable level, yet another source of noise became noticeable – the reliability of the technology needed to sustain system performance. Since much of the concern came from the military side, reliability first became associated with safety and only later with productivity and quality. Continuing this way of explaining the sequence as it happened can perhaps be a basis for thinking about what the next issue, the next source of noise, will be. That will, however, not be attempted here.

The historical reasons for fragmentation are not independent of the psychological reasons in terms of how they were addressed or the attempts to overcome them. Whenever we, as humans, are faced with a problem the intuitive response is to try to solve it as it appears to us. This can be seen as a manifestation of monolithic thinking, the human preference for single and simple explanations that will be described in more detail in Chapter 3. And since a further habit is to solve problems using Depth-Before-Breadth rather than Breadth-Before-Depth – i.e., by searching for the obvious causes without spending too much time trying to understand possible relations to other things – the solutions to each issue were developed on their own. The issues were taken on by different organisations for different reasons or motivations and with different interests or concerns, which led to special interests that soon lost sight of one another.

Because of this fragmentation it has become difficult to see the interconnections or couplings between different issues and therefore to realise that the problems may lie less in each issue than in the ways they are related to each other. In other words, the practical problems may have become larger than strictly necessary because of a failure to recognise the couplings or dependencies in combination with the preference for treating problems in isolation.

Productivity

Productivity is usually defined as the ability to generate, create, or bring forth goods and services that satisfy a human need, but with the essential proviso that the costs needed to produce something must be less than the value of the outcome. The simplest way of expressing that is to define Total Productivity as the ratio between Output Quantity and Input Quantity.

$$Total \; Productivity = \frac{Output \;\; Quantity}{Input \;\; Quantity}$$

Productivity has always been a concern, whether it is producing enough food for a family or population or satisfying market demands for desirable consumer goods. And the drive has always been to increase productivity rather than to simply increase the output volume by increasing the input volume – in other words, to improve the ratio and thereby the efficiency in bringing about the product or service needed.

When work is individual rather than collective people will sooner or later find a way of working that is sufficiently productive for them in the sense that the output volume satisfies whatever the need may be. This may be sensible on a one-by-one basis, where people work for themselves, but as soon as they have to work together – building pyramids, rowing a ship, being soldiers in an army, or manning an assembly line – there may be expectations to productivity and demands that do not tally with the individually acceptable level. This issue became more important after the Industrial Revolution, which required larger capital investments. Those who invested the capital were not necessarily satisfied with the level of productivity that was acceptable to the individual. Moreover, since each individual has a personal standard for productivity, the collaboration may not be as efficient as it could be.

The necessity of economic growth and therefore increased productivity was already addressed by Adam Smith in *The Wealth of Nations* (Smith, 1986; org. 1776). Smith's proposal was that growth is rooted in the increasing division of labour, which meant that larger jobs should be broken down into smaller components. Each worker should then learn to master a smaller part of the work and, by concentrating on that, increase efficiency. According to Adam Smith, productive labour fulfils two important requirements. First, it leads to the production of tangible objects; second, it creates a surplus that can be reinvested into production.

In 1877 Frederick Winslow Taylor started as a clerk in the Midvale Steel Company, one of America's great armour plate production plants, where he by 1880 had advanced to foreman. As foreman, Taylor was "constantly impressed by the failure of his neighbors to produce more than about one-third of a good day's work" (Drury, 1918, p. 23). He was convinced that it was possible to study how work was done in a systematic or "scientific" manner by analysing and optimising workflows and that this could be used to improve efficiency and thereby increase productivity. He observed that most workers who are forced to perform repetitive tasks tend to work at the slowest rate that goes unpunished. Taylor started to put his principles into practice, combining his own method of time studies with Frank and Lillian Gilbreth's work on motion studies, culminating in the publication of *The Principles of Scientific Management* (Taylor, 1911).

The productivity problem was basically solved by applying Adam Smith's suggestion of an increasing division of labour. Taylor's solution was to combine time and motion studies with a rational analysis to uncover the best way of performing any particular task. Taylor also realised that it was important to link each worker's compensation to their output to provide the motivation to change work habits. As he astutely noticed, when workers are paid the same amount they will tend to do the amount of work that the slowest among them does.

Scientific management, or Taylorism, was criticised from the very beginning and is today considered by many outdated, primitive, and denigrating. The purpose here is not to argue for or against Taylorism but simply to point out that it was the publication of *Scientific Management* in 1911 that made productivity an issue and also provided a way to address the problem. *Scientific Management* was able to reduce the noise that came from "unscientific" work practices and individual norms and thereby enhance the signal. The reason it happened at this time is not that the problem was unknown before but that someone actually tried to solve it and succeeded.

Human factors engineering

Scientific Management resolved the productivity issue when work was mostly manual work. But after the IT revolution in the late 1940s, work changed from being work with the body to work with the mind – in other words, from being predominantly manual to being predominantly cognitive. New technologies that increased the speed and precision of industrial processes, whether in manufacturing, construction, or transportation, were assumed to lead to corresponding improvements in productivity. When this did not happen as expected the productivity issue arose again.

The human factor became an issue because humans were seen as too imprecise, variable, and slow. Human performance capacity became a limit to productivity and humans came to be seen as failure-prone and unreliable and hence a weak link in system safety. The problem was clearly stated in a report by the psychologist Paul Fitts from the Ohio State University Research Foundation for the Air Navigation Development Board:

> We begin with a brief analysis of the essential functions of the air-traffic control problem. We then consider the basic question: Which of these functions should be performed by human operators and which by machine elements? . . . Human-engineering research . . . aims to provide principles governing the design of tasks and machines in relation to man's capabilities, and to insure an efficient integration of man and machines for the accomplishment of an overall task.
>
> (Fitts, 1951, p. X)

Where Taylor had analysed work to find the best way of its being carried out by a human worker, Fitts analysed work in terms of the needs of various capabilities to be able to assign specific parts of the task to either humans or technology. Fitts proposed that function should be based on a comparison of humans and machines in terms of sets of specific abilities that have become known as Men-Are-Better-At/Machines-Are-Better-At

(MABA-MABA) lists (Dekker & Woods, 2002). In the short term this seemed to solve the problem of productivity and introduced automation as a near ubiquitous solution. Today the two approaches co-exist and are used together. Fitts's approach not only helped to solve the productivity problem by reducing the noise to an acceptable level, but it was also applied to the safety problem and, in a more limited way, to the quality problem.

Lean manufacturing

Taylor's efforts can in hindsight be seen as an attempt to eliminate the waste that comes from individual levels of aspiration with regard to work. The focus of waste as an impediment to productivity reappeared around 1990 in the guise of lean manufacturing/lean production (or just Lean). The purpose of lean manufacturing is to eliminate everything that does not add value to a process. It is a method that relies on a collaborative team effort to improve performance by systematically removing waste, of which there are eight common kinds: defective products that are unfit for use; over-production; the need to wait, e.g., for materials, thereby delaying the process; waste of human resources and competence; unnecessary transportation of materials, product, people, equipment, or tools; excess inventory of products and materials; unnecessary movement by people; and doing more work than required to complete a task. The similarity with scientific management and Taylor's original motivation to reduce the waste of human resources is easy to see.

Legacy of the production issue

The solutions to the productivity issue, from scientific management via HFE to Lean, can be seen as having created a silo – an approach and a way of thinking that effectively manages production but does it on its own premises (cf. Figure 2.1). This has created a legacy that in itself reinforces the fragmentation. The main parts of this legacy are:

- Task decomposition: The use of decomposition to make it easier to understand something and to manage something successfully was, of course, not new. The aim of scientific management was to analyse tasks to determine most efficient performance. The success in doing this laid the foundation for task decomposition as the universal approach to understand work and human activities. In the 1950s human factors engineering made it the *de facto* standard, first as task analysis and later as cognitive task analysis.
- Specialisation and standardisation: Task decomposition served to identify the basic or elementary steps of a task. People were then selected to achieve the best possible match between task requirements and capabilities and trained to ensure specified performance. Standardisation of work and compliance with the standards were therefore essential.
- Work-as-Imagined: Scientific management did not use this term, but the emphasis on "the one best way" to carry out a task represents the same idea. Actual work situations were studied but only to find out where improvements could be made to what people did rather than to understand the nature of Work-as-Done.
- Elimination of waste: The paramount concern for Taylor was the elimination of waste in the sense that people did not do nearly as much as they could; in other words, their potentials were wasted. Lean production extended and refined this and turned it into a singular management philosophy.

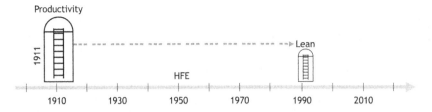

Figure 2.1 The first silo (productivity).

Quality

Scientific management, the concept and the practical methods, demonstrated how productivity could be increased by finding the best way of carrying out a task. This eliminated or dampened a major source of "noise," thereby making it easier to detect the "signal." Removing a major source of noise, however, revealed that there were other sources that so far had gone unnoticed. One of these was deficient quality, which made it impossible to achieve the level of standardisation of components that was required to speed up assembly and hence productivity. The following passage from the preface to the book *Economic Control of Quality of Manufactured Product* introduced quality as the second issue:

> Broadly speaking, the objective of industry is to set up economic ways and means of satisfying human wants and in so doing to reduce everything possible to routines requiring a minimum amount of human effort.
>
> (Shewhart, 1931, p. vii)

The book was written by Walter Shewhart, an American statistician, physicist, and engineer widely known as the father of statistical quality control. The reference to the "minimum amount of human effort" was regrettably not an early attempt to relieve people from numbing routines and dreary work, but rather heralded the growing concern that humans were the variable and potentially unreliable "components" of a well-oiled "machine" that might lead to both production losses and accidents and, in relation to standardisation efforts, to a loss of quality as well.

The cost of low quality

The problem, according to Shewhart, was how to control the production process so that the variability of the quality of the product was acceptably low. This problem was found in processes as diverse as the assembly of telephones or the baking of bread. Variable quality of the output was an impediment for the drive to standardise the production process and make it as efficient as possible. This was so whether the production as a whole took place inside the company or whether parts of it had been outsourced to suppliers. The problem was described as follows:

> How shall the production process for such a complicated mechanism be engineered so as to secure the economies of quantity production and at the same time a finished product with quality characteristics lying within specified tolerances?
>
> (Shewhart, 1931, p. 7)

Variable and unacceptable quality, whether of components or of the final product, was clearly a source of noise that could adversely affect an otherwise smooth production process.

Lack of efficiency in work and production was a problem that had been known for a long time. As mentioned above, Adam Smith had written about it already in *The Wealth of Nations* in 1776, but it must always have been a concern since it is probably a natural human trait not to work harder than absolutely necessary. But quality was a more recent problem and in many ways a consequence of mass production. The initial solution was to use some kind of reference as a basis for comparison in order to decide whether the specific product was of good enough quality. The comparison could be drawings, gauges of various types, or a universal reference such as the metre and the kilogram, introduced by the French Revolution (at least for those who adopted the metric system). This was, however, not practical when mass production was introduced in the U.S. in the 1920s. There was a paramount need to ensure the quality of components and products, or perhaps rather to control the variability of components and products so that it stayed within specified limits. Shewhart's book *Economic Control of Quality of Manufactured Product* clearly described the problem and also provided a practical solution.

Standardisation

Standardisation, to do something the same way time after time, is as old as humankind itself. It is essential for nomadic people who every morning have to pack camp and every evening have to unpack it again – something the Roman armies later developed into a fine art. If this is done in the same systematic way, it saves time and effort. Even something as simple as cooking the daily meal requires standardisation, for hunter-gatherers 20,000 years ago as well as the modern household.

One of the most famous examples of standardisation is Honoré Blanc's demonstration in July 1785 of how a flintlock musket could be built by assembling standardised and therefore interchangeable parts rather than handcrafting the gun – or even the development ten years before by John Wilkinson of Shropshire, England, of a way of making a cannon by boring a hole into a lump of iron so that it was straight and true every single time. Standardisation or uniformity is obviously very valuable, whether it is internalised or externalised.

Standardisation – the criterion for the output – can be internalised when it is work you do yourself and for yourself and when it is work done in the close social group or family. Consider the (wo)man grinding flour to make bread. In this case the standard or criterion is internalised but nevertheless required. If the flour is of uneven quality, baking the bread cannot be done in the same way, and the outcome – the final loaf – may differ in quality. This is so whether we are thinking of the Neolithic Age or the ecological artisan of today. But the moment work is done for someone else, standardisation becomes externalised in the sense that it is someone else's criterion that must be observed. A person can, of course, over time internalise that, but when this happens there is likely to be a change to make it more subjective. Consider Taylor's complaint about workers not putting in a full day. The workers had simply – in one interpretation – developed their own standard for work that was sufficient for them individually and socially but not sufficient for the employer.

When standardisation becomes externalised, explicit control becomes necessary. And to control something, you must understand what it is, what is going on. Standards should only be set on the basis of a good understanding of the work involved. Scientific management did that by studying work (although in a manner that is not considered acceptable or sensible today). Later developments, such as Lean, tend to look at the criterion more than the activities. Lean considers activities in isolation and as Work-as-Imagined rather than as Work-as-Done.

Standardisation straddles productivity and quality but, of course, also relates to safety because the elimination of variability limits the number of situations and conditions that have to be considered in advance, hence the scope of risk assessment. That is why standardisation – in the form of compliance – is often seen as a panacea for all possible hazards.

When quality is achieved through standardisation and uniformity, variability and the potential to adjust performance to the conditions are lost. This will impact safety, productivity, and reliability. Absolute quality, in the sense of through and through standardisation and uniformity, is therefore justified only if quality is the sole concern for a company. But since that is never the case, the advantages of standardisation should always be compared carefully to the disadvantages.

The causes of low quality

Shewhart described the problem as how much the quality of a product could vary and yet still be said to be controlled. Variations in quality, whether in terms of components, completed products, or service functions, were naturally assumed to have an underlying cause or causes. Shewhart's knowledge of statistics made him focus on the distribution of outcomes over time in terms of some quantifiable quality, specifically how large the variability was.

> It seems reasonable to believe that there is an objective state of control, making possible the prediction of quality within limits even though the causes of variability are unknown.
>
> (Shewhart, 1931, p. 34)

The critical question is, of course, how these limits are defined in the sense of how much variability should be controlled and how much left to chance. Shewhart recognised that there was no theoretical basis for doing that and that it was clearly a practical matter. In other words, some variability could be left to chance while some could not. Although it never was stated explicitly, the criterion was (of course) cost in the sense that there is a cost-benefit trade-off in looking for causes. If the variability – or deviation – is very large, then it may be economically efficient or rational to look for the causes in order to do something about them. Indeed, the third postulate in the book stated, "Assignable causes of variation may be found and eliminated" (p. 14). But if the variability is small and the cost of eliminating it is larger than the gain, the variability becomes economically acceptable.

Shewhart proposed that there were two types of causes, assignable and chance, where the determination of assignable causes was statistical. Outcomes that were above the upper or below the lower control limits were by definition said to have assignable causes

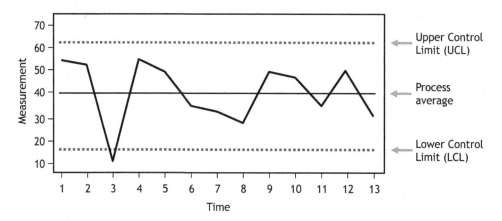

Figure 2.2 Statistical Process Control chart.

(Figure 2.2). This meant that the remaining variability, by definition, would be accept-able, or at least affordable in the sense that the cost of eliminating it would be greater than the benefit.

> In fact, as long as the limits are set so that the probability of falling within the limits is less than unity, we may always expect a certain percentage of observations to fall outside the limits even though the system of causes be constant. In other words, the acceptance of this assumption gives us a right to believe that there is an objective state of control within these limits but in itself does not furnish a practical crite-rion for determining when variations in quality . . . should be left to chance. . . . Furthermore we may say that mathematical statistics as such does not give us the desired criterion.
>
> (Shewhart, 1931, p. 17)

> It shows that when points fall outside the limits, experience indicates that we can find assignable causes, but it does not indicate that when points fall within such limits, we cannot find causes of variability.
>
> (p. 19)

Shewhart's book described in detail how statistical process control could be used to ensure that the variability distribution of output quality stayed within acceptable limits. This established the foundation for quality control and of monitoring a process through sam-pling. Just as Taylor had done, Shewhart presented a theoretically well-founded charac-terisation of the problem and, more importantly, a practical way to solve it. This meant that the noise from unacceptable quality could be brought under control so that the signal – the planned production process and its output – stood stronger.

> By weeding out assignable causes of variability, the manufacturer goes to the feasible limit in adding uniform quality.
>
> (Shewhart, 1931, p. 32)

Assignable and chance causes in socio-technical systems

The unspoken assumption was that process averages could be expected because the processes were well known and well designed – in other words, tractable. The small variations could be explained as due to chance causes, meaning extraneous influences that could not be eliminated by improved design, or that perhaps were not worth the effort. The reason why they happened was thus of no interest. The notion seems to have been that if there were no chance causes, the outcomes – quality – would be perfect; i.e., the causes of the perfect outcomes are deterministic, but the perfect determinism or strict causality may be disturbed by the chance causes. This argument seems to have been reasonable in the 1920s, but it is not today.

Systems and organisations today are recognised as socio-technical systems. These systems are not designed as such but are, rather, intractable (cf. the discussion in Chapter 1). It is conveniently, but wrongly, assumed that things go well because the processes are flawlessly designed and effectively managed, as people perform according to the standards, and so on. Variability due to assignable causes should be eliminated as far as possible; whatever remains is due to chance or common causes but of no practical significance. By contrast, a Safety-II perspective would focus on the common or chance causes since it is they, or rather the variability and adjustments they represent, that bring about the desired outcomes. Resilience engineering and Safety-II both emphasise that acceptable and unacceptable outcomes happen for the same reasons. The real difference between assignable and chance causes is that we make an effort to understand the former but not the latter. But assignable and chance causes differ only in their consequences (phenomenology), not their ontology or aetiology (Hollnagel, 2014). Since the absence of variation – or rather the absence of unacceptable outcomes – is the purpose of the system, we need to understand how this happens. It therefore makes sense to look at what produces the acceptable outcomes and to understand the variability ascribed to chance causes, accepting that unacceptable outcomes happen in the same way.

Legacy of the quality issue

The solutions to the quality issue created a second silo. Shewhart's solution to the problem of unacceptable quality was quickly adopted by industries because it was both theoretically well-founded and very practical. But as in the case of scientific management it did it on its own premises (cf. Figure 2.3) and thereby created a legacy that was

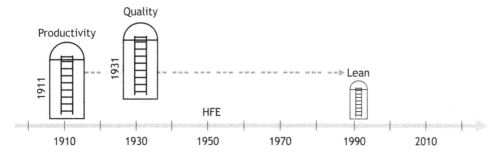

Figure 2.3 The second silo (quality).

added to but not integrated with the legacy from productivity. The main parts of the quality legacy are:

- Statistical process control: Rather than studying the details of how something was done to find the one best way, Shewhart introduced the use of statistical methods to monitor and control a process. The practice of using data from a representative sample to infer something about the larger whole proved itself very effective.
- Assignable causes and chance causes: The distinction between assignable and chance causes offered a way to deal with the statistical outliers but also meant that the common variability was seen as insignificant and – more importantly – uninteresting. This reinforced the common notion of linear causality in the same way that safety management did, although independently.
- Change management as a cycle: Improvements to quality were made in small steps of limited scope rather than by a single, large revision or redesign of a work process. The steps were repeated or iterated in a cycle to produce gradual changes for the better. This became the basis for iterative management approaches such as the PDSA cycle, of which more will be said in Chapter 5.

Safety

Removing a major source of noise by taming productivity revealed that there were other types of noise. One of these was insufficient safety. Heinrich's book *Industrial Accident Prevention* did for safety what Shewhart's book did for quality and indirectly also helped productivity by providing a model for how accidents happened and how they could be prevented.

Safety and quality both became recognised as major issues from the 1920s onwards, as already described, and in both cases the determining event was the publication of a book in 1931. It is tempting to see this as more than a coincidence, although it probably is not. The common background was that the problem with noise from the variability in productivity had been solved. Productivity had been improved using scientific management principles, but it soon turned out that there were other kinds of noise that affected work output and, therefore, productivity. One problem was unacceptable quality, as described above, the other the occurrence of accidents. For an industry accidents could mean disturbances and, at worst, disruptions in productivity as well as unwanted costs. In most cases the costs were recovered from insurance companies, which therefore were very motivated to find ways to improve safety and thereby reduce the number of accidents as well as their severity – which, of course, would increase the "productivity" of the insurance companies themselves. In hindsight it is interesting that quality and safety were both identified as a likely source of noise and both treated it by itself, with little if any appreciation that there might be other factors and awkward mutual dependencies at play. Focusing on the primary concern (safety or quality) was initially successful because the systems at that time were still reasonably tractable.

The cost and causes of accidents

Just as Taylor was concerned that productivity was below what it should be because workers did not work in an "optimal" manner, so Heinrich – and with him many others – was concerned that the lack of safety led to reduced productivity. In the book that marked the beginning of safety as a fundamental concern for the industrial society, *Industrial Accident Prevention: A Scientific Approach* (Heinrich, 1931), Heinrich argued that accidents had a direct as well as incidental cost and that the incidental cost was four times greater than the direct cost. This was later revised by Frank Bird (1974), who in the famous – or perhaps notorious – iceberg model of accident cost proposed a ratio of six to one. Regardless of what the ratio is postulated to be, the fact remains that accidents can be expensive and have an impact on productivity in many different ways. Heinrich even made the claim that "a 'safe' factory is eleven times more likely to be 'productive' than is an 'unsafe' factory" (Ibid., p. 33). In another article he argued that "the real causes of accidents are likewise the real causes of decreases in efficiency, production, and profits. In short they denote conditions that are morally and economically improper" (Heinrich, 1929, p. 5).

Because accidents are costly and thereby directly and indirectly affect efficiency and productivity, there is a strong need to prevent them from happening. This has for obvious reasons been a concern for humans and societies for millennia, and people have forever tried their best. Reports of fatal accidents at work are known as far back as 1540; the HM Factory Inspectorate was established in the United Kingdom in 1833, and the American Society of Safety Engineers was founded in the USA in 1911 after the Triangle Shirtwaist Factory fire. But as Heinrich pointed out, there had so far been no systematic study of why accidents happened:

> In an age of exact knowledge (such as this in which we live today) when facts of proved value are replacing theories and when business, under the pressure of economic necessity, must concentrate upon the things that count, it is startling indeed to find that in at least one respect the efforts of thousands of individuals are misdirected and that the attainment of a worthwhile objective is seriously delayed through failure to recognize an obvious truth. The "objective" referred to is the control of industrial accidents. The "obvious truth" is that the reduction of accidents can only come about through knowledge of basic accident causes.
>
> (Heinrich, 1928, p. 121)

The solution

Just as Taylor had argued for a scientific study of work and proposed scientific management theory as the solution, so Heinrich argued that accidents should be studied scientifically:

> The prevention of accidents is a science, but it is not so recognized nor is it treated scientifically today. In our endeavors to solve other problems we proceed logically enough to find out what is wrong and then to make it right; but when it comes to accidents the same degree of logic is not so evident.
>
> (Ibid.)

This is consequently what he set out to do in his book from 1931, where the second part of the title significantly was *A Scientific Approach*. The scientific approach was the theory of accident causation that later became famous as the domino model. In its most simple form the reasoning proposed a sequence of causes and effects where the cause would lead to an accident, which then would lead to an injury. In consequence of that it was possible to find the cause of the accident by reasoning backwards starting with the injury. This is the basis for the still widely used method known as root cause analysis. (As an interesting aside, Heinrich believed that psychology lay at the root of the sequence of accident causes – perhaps a foreshadowing of the human factors engineering that was yet to come?) Given that accidents could be studied and understood in this way, the solution was equally simple: find the root cause and eliminate it.

As with scientific management this approach was immediately successful, to the extent that it has practically become second nature in how we think about safety. So much has been written about the domino theory, linear causality, zero accident programmes, etc., that there is no need to continue that discussion here. Regardless of whether the domino model is sensible today, the fact remains that the concern for safety came on to the scene in 1931, and the reasons were the same as for productivity and quality. Someone described the problems succinctly and also offered a solution that worked – at least in the short term.

Safety-I and Safety-II

The understanding of safety as being a condition where as little as possible goes wrong, where there are as few accidents as possible, where the possibility of harm to persons or property damage is reduced to and maintained at an acceptable level, or any other definition along the same lines, quickly became the established truth. This is, of course, sensible since everyone would want to avoid harm or injury to oneself as well as others. But the focus on accidents and injuries means that safety in this way is defined more by its absence than by its presence, as pointed out by James Reason (2000). Safety management tries to achieve its goal by means of elimination, prevention, and protection, and safety is therefore measured by something that gets smaller and smaller until it – hopefully – becomes zero. This has turned out to be problematic and in practice impossible. A major reason is that work – and societies – has become increasingly difficult to understand as described in Chapter 1. Because the usefulness of the tried and tested approaches began to dwindle during the 1980s and 1990s, new ways of thinking about safety started to emerge. Around 2000 the concept of resilience found its way into discussions of safety, and soon after the first book on resilience engineering was published (Hollnagel, Woods, & Leveson, 2006). The contrast between the established way of interpreting safety and the alternative provided by resilience engineering eventually became expressed as the difference between Safety-I and Safety-II (Hollnagel, 2014). Safety-I and Safety-II agree on the ultimate aim – namely, that adverse outcomes should be avoided as far as possible – but differ with regard to how this can best be achieved. The approach taken by Safety-I is protective and tries to make sure that as little as possible goes wrong. The approach taken by Safety-II is productive and tries to make as much as possible go well, the simple logic being that if something goes well, then it simultaneously cannot fail. Therefore, the more things that go well, the fewer accidents there will be.

Legacy of the safety issue

The solutions to the safety issue created a third silo (Figure 2.4). Heinrich's work actually developed over a number of years, and the book went through several editions, the fourth edition published in 1959 (a fifth edition was published in 1980, but Heinrich passed away in 1962). The domino model and root cause analysis quickly became the *de facto* standard in accident analysis and in many ways remain so today. Heinrich's practical take on how to deal with industrial accidents created a legacy that was added to but not

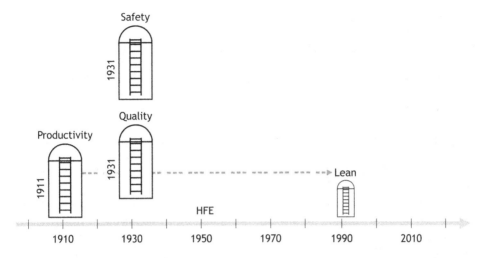

Figure 2.4 The third silo (safety).

integrated with the legacies from productivity and quality. The main parts of the safety legacy are:

- Linear causality and root cause analysis: The linear cause-effect thinking implied by the domino model provided a simple but effective principle for dealing with safety problems: find the root cause and eliminate it. The conclusion that (all) accidents are preventable has become the basis for the zero accident myth that continues to this very day.
- Human error: The penultimate domino in the original version of the domino model was "fault of person," preceded by "social environment, ancestry." This introduced the human error as an important cause but without the elaborate models of "human error mechanisms" that later flourished.
- The hidden cost of accidents: The demonstration that accidents not only had direct costs but also hidden or indirect costs was important in widening the concern for safety from the primary and immediate effects to the secondary effects that may have consequences in the longer term, for the organisation as well as the people at work.

Altogether Heinrich's work can now be seen as having provided the foundation for Safety-I, although that label was not applied until seventy years later.

Reliability

The fourth issue, or source of noise, was reliability. The consequences of insufficient reliability slowly became visible when the problems of productivity, quality, and safety had been solved, at least for the time being. A major transition happened when computing machinery and information technology became widely used, particularly after integrated circuits were invented in 1958. In the beginning of the 20th century work was mainly manual and on a macroscopic level, hence the value of time and motion studies. As new technologies were developed during and after the Second World War, work changed from being manual to being cognitive, from being hands on to being more and more control and monitoring. The rapidly growing use of computers and information technology led to an increasing dependence on the functions and performance of technologies that were not fully mastered or controlled. This started the vicious circle that continues today and that is becoming worse and worse. Because of that work environments quickly became so complicated that they were vulnerable to problems of reliability.

Reliability is technically defined as the probability of success. Therefore, if the probability of failure is negligible or even zero, reliability is high or even certain. This corresponds well to the common, non-technical understanding of reliability as the quality of being trustworthy or of performing consistently well. A good illustration of this is the story of the high-pressure steam engine, patented by James Watt in 1769. In the beginning of the 19th century it came into widespread use in steamboats and industrial production (Leveson, 1992). But the high-pressure steam engine was an unreliable component since it could easily explode and thereby injure or kill crew, passengers, and workers. The U.S. Commissioner of Patents estimated that between 1816 and 1848 there had been 233 steamboat explosions, resulting in 2,562 persons killed and 2,097 injured. This situation changed with the introduction of the low-pressure steam engine, which was a far more reliable component. But even for highly reliable components, the possibility that they may fail exists and is seen as a risk.

The term *reliability* is often used to characterise technology, as when a car is said to be reliable. But it can also be used to talk about people, as when a person is said to be reliable, and even to characterise organisations. Reliability is also commonly used to describe the consistency of how something is done – for instance the reliability of a test or a measurement. Reliability was from the beginning associated with safety (risk) rather than productivity since the malfunctioning of components could result in accidents and harm. In relation to the use of technology, reliability was defined by the U.S. military in the 1940s to mean that a product or piece of equipment would operate as required and for a specified period of time. In this way reliability became linked to the ability – or inability – to achieve a desired objective. This was an issue of obvious concern for the military, but it is also of great concern for all uses of technology, in industry and in private use.

As mentioned earlier, in 1950 the Advisory Group on Reliability of Electronic Equipment (AGREE) was formed by the U.S. Department of Defense to investigate reliability methods for military equipment. Just as in the cases of productivity, quality, and safety, the group proposed a way to solve or overcome the reliability problem. In this case it was not a new method or approach as such, but a recommendation of three lines of activity that together addressed the core of the problem (Saleh & Marais, 2006). These were (1) to systematically collect data from the field and use these to identify causes of component failure, (2) to issue quantitative reliability requirements to suppliers as a contractual obligation, and (3) to develop ways to estimate and predict component reliability before equipment was built and tested. A result of these activities was the AGREE report published in 1957, which provided the assurance that reliability could be specified, allocated, and demonstrated. The AGREE report can therefore be seen as the point where there was a formal recognition of reliability as a problem or a source of noise.

There is another, perhaps even more important, way in which reliability plays a role. In this case the problem is not whether the malfunctioning of a component can end in an untoward or adverse outcome, accident, or incident, but rather whether the functioning of a component or subsystem is reliable in the sense that it will not disrupt the smooth operation of the system (a factory or a production line, for instance). In other words, can we rely on or trust something – or someone – to deliver what is required for as long as it is needed. Reliability in this sense became necessary when some parts of the production (supplies, etc.) were delegated to others and hence no longer under direct control. In this case it is important to have a reliable source. This becomes even more extreme when the source is something like electrical power, water, communication, or supply lines in just-in-time production, etc.

In the cases of productivity, quality, and safety it had been possible to propose a simple way to solve the problems. In the case of reliability the much weaker solution was to offer a combination of calculation and specification, which in turn requires quality. Quality and reliability are therefore closely linked even though they are separate chronologically. The reliability problems furthermore turned out not to be limited to technical components; they soon also came to include humans and organisations.

Human reliability

The problems with reliability were from the beginning linked to technology, on both the component and system levels, and it was assumed that it would be sufficient to address these. This optimistic view changed dramatically with the accident at the Three Mile Island nuclear power plant in 1979. In searching for a cause that could explain how the accident had happened, it became clear that the reliability assessment of the installation had missed one essential "component" – namely, the human operator. While mechanical or technical failures contributed to the accident, the operators also failed to correctly recognise the situation as a loss-of-coolant event. This in turn was due to a combination of inadequate training and substandard human factors design of the user interface and hence not really an individual "human error" as such. In the understanding of how the accident happened, the operators' performance was seen as being unreliable, much in the same way that a component could unreliable. Humans generally came to be seen as fallible machines for which it was necessary to be able to assess their reliability. (Unfortunately, the solutions recommended for technical reliability issues could not simply be transferred to human reliability; for instance, it was not possible to issue reliability requirements to the supplier.) The response that followed the TMI accident consolidated previous efforts that since the late 1940es had tried to address the problems of the reliability of human performance, resulting in the establishment of the field of human reliability assessment, which grew rapidly during the 1980s and 1990s.

High reliability organisations

The American professor of sociology Charles Perrow published a book in 1984 with the provocative title *Normal Accidents: Living with High-Risk Technologies*. In the book Perrow analysed the social side of technological risk and argued that the engineering approach to ensuring safety did not work because it led to increasingly complex systems where failures were inevitable. Others soon pointed out that there were actually many organisations that seemed to be able to avoid the expected and dreaded "normal" accidents. They argued that system accidents were manageable rather than inevitable, at least in some cases. Organisations that were able to function in an essentially error-free manner were called high reliability organisations (HRO) (Roberts, 1989).

In HRO, reliability is still understood in the sense of trust, that you can rely on or trust an organisation to perform in a reliable way, that it will operate as required when needed. In the wider interpretation, reliability therefore does not just mean that the likelihood of failure is low – or lower than "normal" – but rather that the likelihood of performing as required and expected is high enough to be unproblematic. This interpretation is very close to the meaning of resilience as it is used in resilience engineering (Hollnagel, Woods & Leveson, 2006), and it is therefore not surprising that researchers from the HRO community were among the first to use the term *resilience* (Weick & Sutcliffe, 2001).

If the solution proposed for the problems arising from insufficient reliability of technical equipment was less concrete than the solutions proposed for productivity, quality, and safety, then matters hardly improved for human reliability or organisational reliability. For human reliability a number of methods were developed to calculate the probability of a "human error" since that was needed to fit into the engineering models. But little was done in terms of actually suggesting ways to improve human reliability, except perhaps by replacing humans with automation (as in the good old days of human factors). In the case of organisational reliability or high reliability organisations, five characteristics were proposed for HROs: preoccupation with failure, reluctance to simplify interpretations, sensitivity to operations, commitment to resilience, and deference to expertise. But little was done in terms of showing how these characteristics could be determined or measured, and even less was made concrete in terms of how they could be managed and improved.

Legacy of the reliability issue

The solutions to the reliability issue created a fourth silo (Figure 2.5), and perhaps also a fifth and a sixth, if human reliability and organisation reliability are considered separate issues. Human reliability clearly to some extent is, while organisational reliability remains an uneasy mélange of HRO and safety culture. The reliability issue was never as distinct as the three previous, and it could not present a straightforward and easily understood – even if somewhat illusory – solution. Productivity can be measured, quality can be measured,

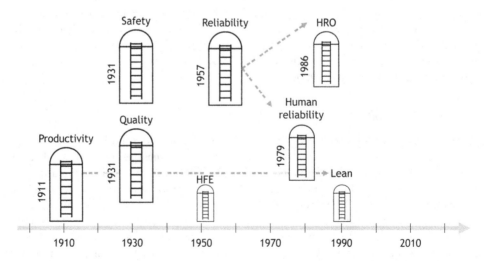

Figure 2.5 The fourth silo (reliability).

and even safety can be measured (although it is measured by its absence rather than its presence), but reliability can only be calculated. This in no way detracts from its importance, but it makes it less tangible in day-to-day management. Reliability analyses superficially relied on many of the techniques used by quality and safety. The main legacy of reliability is therefore this:

Probabilistic reasoning:

* The focus of reliability was whether something (at first technology, but later also humans and organisations) would fail or malfunction. This could not be approached in a deterministic sense but only by describing how probable it was that something would happen. This opened the door for the probabilistic safety analyses of large systems that today have become indispensable. This type of reasoning perpetuated the legacy of decomposition from the productivity issue and linear causality from the safety issue, although without any actual conceptual integration.

The common legacy

The fragmentation came about because the solutions developed for each of the four problems reinforced the belief that problems could be solved by isolating them, by finding their causes, and by dealing with these. It was perhaps not so surprising in the case of productivity since it appeared on the scene first. Historically, Taylor's scientific management was nevertheless soon sidetracked by Shewhart's concern for quality on the one side and Heinrich's concern for injury and accidents on the other. For both quality and safety the economic consequences dominated and were in fact the main motivation. Despite that, it was never seriously attempted to relate either quality and productivity or safety and productivity on the conceptual level, i.e., to consider whether the issues were coupled or how they might depend on each other. This is in some sense even more puzzling in the case of reliability since it went from technical reliability to human reliability to organisational reliability. Yet for each domain reliability was used as a symptomatic remedy without really going into depth. In all cases the legacy of linear cause-effect reasoning was demonstrated to work, and it was therefore reinforced to such an extent that by now it is simply taken for granted. But linear causality is not adequate for systems and organisations that are non-linear and intractable, which many are today. Part of the reason for the fragmentation lies in how present-day technological societies have developed and matured. But another intrinsically related reason can be found in how humans reason and think, in the psychological reasons for fragmentation, to which Chapter 3 is devoted.

3 Psychological reasons for the fragmented view

Introduction

There are two major reasons for the fragmented view that dominates how we manage organisations. The historical reasons, the fact that the various issues became recognised at different points in time and were addressed one by one, have been described in Chapter 2. What the psychological reasons mean and which effects they have will be explained in this chapter. The main argument is that the fragmented view, to a considerable extent, is a consequence of how our minds work, of how we think about the world. (By the world, I mean the artefacts, systems, and organisations that surround us and that we are part of, most of which we have built to provide functions and services we deem necessary for our existence, well-being, and survival and many of which we have made ourselves dependent on – often without realising it before it was too late.) In short, the psychological reasons are the ways in which people willy-nilly think about things and therefore how they explain them, interact with them, and manage them. It seems to be part of human nature to do this in as simple a way as possible even when it – in hindsight as well as in foresight – is rather obvious that it does not work very well.

It might be tempting to see the psychological reasons as synonymous with the human factor in the sense of the characteristics of the "people at work." Yet to do so would miss the essence of the argument. Quite apart from the fact that the "human factor," in many cases, is used mainly as a convenient way of explaining why something fails or does not work as well as it should, the psychological reasons simply cannot be attributed to others and hence nicely be disposed of as what somebody else does. The psychological reasons apply to all of us, including I who write these words and you who read them.

Philosophy, psychology, and even human factors (as a scientific discipline) have all tried to understand and explain how the human mind works. There are countless books and papers about this as well as a great number of theories, models, ideas, and speculations – not all of them consistent and some of them contradictory. There is as yet no unified theory of mind and considerable disagreement about whether that would make sense at all since psychology is neither a formal nor a natural science. The purpose of this chapter is not to get into an argument about how the mind works, but rather to simply point out that there are some fundamental phenomena in how humans think that should be recognised because they have consequences for what people do and how they do it. To prevent unnecessary confusion the focus is on a description of how the mind works, not how the brain works. The description will be limited to a small number of omnipresent characteristics of human thinking and reasoning. It is, of course, somewhat ironic that the fragmented view also applies to how it can be described, hence the understanding of how the mind works. Without reflecting on and acknowledging that, analyses and explanations will forever remain trapped in established and constraining mindsets.

Limited span of attention

The natural place to start is to look at characteristics that for better or worse may be called biological or biologically based. This means that they are the consequence of something, described variously as how the brain is wired or how human neurophysiological processes take place. Quite apart from the fact that the scientific knowledge about these processes is incomplete, it is impossible to do something about them, as they are beyond any means of systematic control. Unquestionably, the most important characteristic is the limitation of attention, also called the limited span of attention. To pay attention to something means that it is taken notice of, either because it is seen as interesting or important because it is a strong signal that stands out against the background noise or because it is unusual or unexpected. The span of attention, moreover, has two different meanings. One is the length of time it is possible to pay attention to something, to keep it in focus or to concentrate on it. (It is common to lament the decline in span of attention ascribed to modern technology. According to a study by Microsoft, the average human being in 2019 had a span of attention of eight seconds, a sharp decrease from the average span of attention of twelve seconds in the year 2000. In comparison, a goldfish is said to have a span of attention of nine seconds.) The other is the number of things or thoughts that can be kept in mind or paid attention to at the same time. In modern psychological terminology this is often referred to as the span of short-term memory or as the limits on the capacity to receive and process information.

William James, an influential American philosopher and psychologist, wrote the following about attention:

> The number of things we may attend to is altogether indefinite, depending of the individual intellect, on the form of the apprehension, and on what the things are. When comprehended conceptually as a connected system, their number may be very large. But however numerous the things, they can only be known in a single pulse of consciousness for which they form one complex "object" (p. 275ff.), so that properly speaking there is before the mind at no time a plurality of ideas, properly so called.
> (James, 1890, p. 405)

There is in principle no limit to the number of things a person may attend to at one time or another, which basically means that there is no limit to the number of things a person can consider and know something about given enough time. But there is a rather severe limit on *how many* things or *how much* we can attend to at the same time. According to William James, the number is just one; i.e., it is possible to concentrate on or be conscious of only one thing at a time. There is also a limit on *how long* this can be done without losing attention, before the mind begins to wander, although this limit is less well defined, being both person- and situation-dependent. In both cases the reason is presumably something that is intrinsically associated with how the brain works, which means that it is a biological limitation that simply must be accepted. Although humans have invented a number of ways to overcome this limitation in practice, the fact remains that it is possible to attend to what happens only as fragments, not as a whole.

The problem of the limited attention span has been studied extensively since the early days of experimental psychology in the 1860s. It was also the topic of what is arguably one of the most influential papers in contemporary psychology. Written by George A. Miller (1956), the title was *The Magical Number Seven, Plus or Minus Two: Some Limits on Our Capacity for Processing Information*, which indeed by itself attracts attention. The paper took a closer look at the limited span of attention by means of the then novel concept of human information processing. Miller analysed a wide range of experimental findings and came to the conclusion that there is a limitation in the span of attention, now also called short-term memory, famously labelled the "magical" – but also apocryphal – number seven. (The apparent clash with James's estimate is because he and Miller referred to two different phenomena; James wrote about how much someone can be conscious of, while Miller wrote about how much someone could hold in mind for a short period of time, such as a phone number while dialling.) Regardless of whether the number is seven or something less, or even whether it makes sense to propose a number, the fact remains that the human span of attention is limited. This obviously affects the ability to describe, understand, and interact with the world. The limitation did not present a serious problem in the days of yore, such as towards the end of the 19th century, both because there was little need to attend to something for extended periods of time and because work was predominantly manual and hence based on perceptual-motor rather than cognitive skills. But the same limitation is a problem now and has increasingly been so since the information technology revolution in the 1950s.

Another difference between James and Miller is that James wrote about how someone perceives the world and how attention limits the number of things a person can evoke when trying to think about something or perhaps make a decision. In either case this was something done at leisure, so to speak, rather than as a part of working. Miller was more concerned with the need to discriminate among multiple items and the need to remember things as part of doing something – in other words, the consequences of the limited span of attention on the ability to do something in practice. The result is in both cases a fragmented view, but it becomes a more serious problem when it is part of doing something, such as managing an organisation.

The limited span of attention is reflected in the many systems or frameworks that have been developed to help make sense of the confusing reality, each offering a solution by proposing a small set of discrete categories, thereby unintentionally reinforcing the fragmentation they intended to overcome. (It is also interesting to note that nearly all of them comply with the "magical" number seven.) The best known version of these is probably the abstraction hierarchy (Rasmussen, 1986), but there are others such as James Miller's living systems theory (1978), Stafford Beer's viable systems model (Beer, 1984), and David Snowden's Cynefin framework (Kurtz & Snowden, 2003), to name but a few.

Information input overload

George Miller's research took place in the context of the scientific and technological developments that became the basis for the now indispensable information technology. Computing and display technologies made it possible to gather, transmit, and present more information faster than at any time before. In relation to this, a crucial consequence of the limited span of attention is the ability – or perhaps rather the inability – to attend to and make use of the information that can be put in front of people or pushed onto them. The technical term for an excessive quantity of data is *information input overload* (IIO). The problem, by the way, is far from new, as the writer of Ecclesiastes 12:12 complained that "of making books there is no end." Yet as long as work was mainly manual, such as when scientific management came into existence, this was of limited concern. Work at that time was not as fast-paced as it is today. But as work gradually but inevitably changed from being mainly manual to being mainly cognitive, or in other words to becoming work with the mind rather than work with the body (or hands), the ability to deal with information in the work situation became crucial. At the same time the quantity of information changed from being limited by the number of gauges and instruments that could be fitted onto a physical space to becoming nearly unlimited as the computer display became a narrow keyhole through which the vast data space within could be accessed (Woods & Watts, 1997). Twenty years ago this was recognised as a problem for human-machine systems in work contexts; today it is a problem for us all.

Whereas the limited span of attention is an absolute condition with a biological basis, IIO is more of a relative condition. Information input overload means that the momentary capacity to deal with or process the information that is being presented is insufficient. Information input overload can be a result of an increased rate or amount of input information (things happening too fast or too many things happening at the same time), of a temporarily reduced capacity to process that information (fatigue, stress, interruptions, competing tasks), or, of course, both simultaneously. The unavoidable consequence of IIO is that it becomes impossible to take in or pay attention to all the information available, which means that something inevitably will be missing. Since the condition is one that happens regularly, humans have over time developed a way of coping or responding, ranging from the temporary non-processing of input to abandoning the task completely. In between these extremes are a number of different strategies that on the whole aim to preserve the essential information while at the same time allowing work to continue and avoiding delays that may lead to a deterioration of the situation. The reactions can thus be seen as an example of a trade-off between efficiency and thoroughness, described in more detail later. Defining what the essential information is cannot, of course, be done objectively but must depend on a person's current interpretation of the situation and the current goals. The very existence of these reactions therefore implies a rather tight coupling between the information that is needed in a situation and how this information is treated. The basic modes of reaction are shown in Table 3.1.

Regardless of how a person responds, the inevitable result is a fragmented view or understanding. IIO was initially mostly studied in situations where an individual was faced with too much information at work, although that today seems to apply to most of us throughout our waking hours. A theme stressed in the literature is the paradoxical situation where, despite an abundance of available information, it can sometimes be difficult to find the relevant information when it is needed. The condition of insufficient information has been called information input underload (Reason, 1988). (An unintended result of coping with IIO may be a situation of information input underload because the "wrong" information has been filtered or deselected.) This is unquestionably also an issue in how organisations are managed. Here the ability to gather information from everything and at any time uncritically is generally accepted as an advantage – for instance in the case of network-centric organisations where every part of a system is a data collection node. While getting more information may have advantages it also has the serious disadvantage of leading to IIO. A popular "solution" is the use of a management dashboard to present all important information in a single place to allow fast decisions based on the latest data. In practice this is more likely to make the fragmentation even larger.

Table 3.1 Coping strategies for information input overload (IIO)

IIO strategy	Definition	Criterion for use
Omit	Temporary, arbitrary non-processing of information; some input is lost.	It is important to complete the task without further disturbances.
Reduce precision	Trading precision for speed and time; all input is considered but only superficially; reasoning is shallower.	It is important to reduce or compress time but not miss essential information.
Queue	Delaying response during high load on the assumption that it will be possible to catch up later (stacking input).	It is important not to miss any information (this is efficient only for temporary conditions).
Filter	Neglecting to process certain categories; non-processed information is lost.	Time/capacity restrictions are really severe and it is sufficient to note only large variations.
Cut categories	Reducing level of discrimination; using fewer grades or categories to describe input.	
Decentralisation	Distributing processing if possible; calling in assistance.	Availability of additional resources
Escape	Abandoning the task; giving up completely; leaving the field.	System self-preservation

Although IIO may seem like a particularly modern phenomenon, not least because of the ubiquitous use of and addiction to information technology, the fundamental categories defined in Table 3.2 can be found in a paper by James Grier Miller from 1960. At that time only a few were able to imagine the impact of the personal computer, and, for comparison, the World Wide Web was not invented until nearly 30 years later in 1989.

Bounded rationality

While the limitations in the span of attention have a biological basis and therefore are impossible to avoid, the typical responses to information input overload have a biological and a cognitive basis. If a limitation has a cognitive basis it means that it may be possible to overcome it even though it could require a considerable – and perhaps excessive – level of effort. The IIO responses are stereotypical patterns, some of which are instinctive, while others are deliberate – like omitting information or queueing. The IIO responses thus straddle the biological and cognitive domains. The advantage is that they bring temporary relief from the overload conditions. The disadvantage is that they result in a fragmented and incomplete understanding of the situation and may even lead to a condition of information input underload. This may have many consequences, one of the more conspicuous ones being on how decisions are made.

Rational thinking, or rather systematically correct reasoning, has been a human ideal at least since the time of Aristotle, who in the *Organon* formulated the principles for the formal accuracy of arguments. Although a formally correct argument does not guarantee that the conclusion – or decision – is factually correct, it at least seems unlikely that something systematically can be factually correct if it is not also formally correct. In other words, formally correct reasoning appears to be a prerequisite for factually correct conclusions. This is borne out by a contemporary definition of rationality as the quality of being consistent with or based on logic in doing something, presumably reasoning or decision-making.

In classical or normative decision theory, and specifically the part of decision theory that has to do with economic behaviour, reference is often made to an ideal decision maker known as *homo economicus*, or the rational economic person. This decision maker is assumed to be completely informed and to know all courses of action as well as the outcomes of any action; this decision maker is also assumed to be infinitely sensitive – which means that the characterisation of alternatives is infinitely divisible – and, finally, rational. The latter term implies that the decision maker must be able to put the alternatives into a weak ordering, to determine preferences among alternatives, and to choose in order to maximise something – utility, risk, etc. Although palpably wrong, the myth of *homo economicus* is nevertheless important because it defines the perfectly rational decision maker and hence the criteria for what an optimal decision is. It assumes that agents always act in a way that maximises utility as a consumer and profit as a producer and, more importantly, that they are capable of arbitrarily complex deductions towards that end. They will, as rational decision makers, always be capable of thinking through all possible outcomes and choose that course of action that gives the best possible result.

One consequence of psychologically based fragmentation is that decisions can never be rational in the sense described above. This has obviously been recognised both by decision theory and management theory. Decision theory has tried to develop more "humane" versions of rationality that, on the one hand, would meet some of the defining characteristics while, on the other, also being psychologically realistic. Alternative views have also been proposed, such as descriptive decision-making and naturalistic decision-making (Klein et al., 1993). Management theory has on its side challenged the orthodoxy of the rational decision maker and instead proposed alternative descriptions. The best known of these are the characterisation of *satisficing* by Simon (1956) and the "muddling through" description of decision-making by Lindblom (1959). In both cases the decision-making is seen as going through the following stages: (1) define the principal objective, (2) outline a few obvious alternatives, (3) select an alternative that is a reasonable compromise between means and values, and (4) repeat the procedure if the result is unsatisfactory or if the situation changes too much.

Simon argued that *satisficing* was inevitable because of the limitations of human cognition and hence was biologically based. He used the term *bounded rationality* and emphasised that rationality was limited by the tractability of the decision problem, by the cognitive limitations of the mind, and by the time available to make the decision. Decision makers, in this view, act as satisficers, seeking a satisfactory solution rather than an optimal one. Lindblom took a more pragmatic approach and argued that muddling through was simply the best – and perhaps only realistic – way of making decisions in-between the economic and the bounded rationality. Muddling through may therefore be seen as having a social rather than cognitive basis. If a limitation has a social basis it means that it, in principle, can be overcome but doing so would go against the mainstream and therefore continually require explanations and justifications. It is therefore usually more efficient to just go along. Regardless of whether a limitation has a biological, cognitive, or social basis, the result is a fragmented view because only part of the available information is taken into account when a decision is made. This will obviously have consequences for how an organisation is managed, for the quality or correctness of expectations (anticipations), and for planning on a strategic and a tactical level.

The *satisficing* strategy is a good example of the use of heuristics in decision-making. Heuristics are simple but practical strategies to form judgements and make decisions that involve focusing on the most relevant aspects of a complex problem. A heuristic is not optimal, perfect, or rational but is usually sufficient to reach an immediate, short-term goal. While the heuristics do not compensate for fragmentation as such, they go some way towards alleviating the consequences. Some of the better known heuristics are representativeness, where A is seen as representative of B by the degree to which A resembles B; availability, where the likelihood of something is based on the ease with which a particular idea can be brought to mind; and anchoring, where different starting points lead to different estimates that furthermore are insufficiently adjusted or corrected (Tversky & Kahneman, 1974).

Efficiency-thoroughness trade-off

It is a common condition of human work, whether individual or collective, that the resources to do something often, if not always, fall short of what is required. The most frequent shortcoming is a lack of time, but other resources such as information, materials, tools, energy, and manpower may also be limited. (Information may, for instance, be missing as a result of coping with IIO.) People nevertheless usually manage to meet the requirements of acceptable performance by adjusting what they do to meet the demands and the current conditions – or, in other words, to find a balance between demands and resources. This ability to adjust performance to match the conditions can be described as an efficiency-thoroughness trade-off or ETTO (Hollnagel, 2009).

- Efficiency means that the level of investment or amount of resources used or needed to achieve a stated goal or objective is acceptable. The resources may be expressed in terms of time, materials, money, psychological effort (workload), physical effort (fatigue), manpower (number of people), etc. The acceptable level or amount is determined by a subjective evaluation of what is sufficient to achieve the goal, i.e., good enough to be acceptable by whatever criterion is applied locally as well as by external requirements and demands. For individuals, the decision about how much effort to spend is usually not conscious, but rather a result of habit, social norms, and established practice. For organisations, it is more likely to be the result of a direct consideration, such as a lean management philosophy, although this choice in itself will also be subject to the ETTO principle.
- Thoroughness means that an activity is carried out only if the individual or organisation is confident that the necessary and sufficient conditions are in place so that the activity can achieve its objective and that it will not create any unwanted side effects. These conditions comprise time, information, materials, energy, competence, tools, etc. The confidence is very much subjective and the threshold may easily be subject to various forms of social pressure. More formally, thoroughness means that the pre-conditions for an activity are in place and that the execution or operating conditions can be established and maintained so that the outcome(s) will be the intended one(s).

The efficiency-thoroughness trade-off (ETTO) principle recognises the fact that people (and organisations), as part of their activities, frequently – if not always – must make a trade-off between the resources (primarily time and effort) they spend on preparing to do something and the resources (primarily time and effort) they spend on doing it. The trade-off may favour thoroughness over efficiency if quality and safety truly are the primary concerns and efficiency over thoroughness if throughput and output are the primary concerns. It follows from the ETTO principle that it is never possible to maximise efficiency and thoroughness at the same time. By contrast, an activity cannot expect to succeed if there is not a minimum of either.

Any specific efficiency-thoroughness trade-off – sometimes referred to as an ETTO rule – is, of course, a heuristic. In practice, people use shortcuts, heuristics, and rationalisations to make their work, problem-solving, and decision-making more efficient. At the individual level, there are many ETTO rules – e.g., "It will be checked later by someone else," "It has been checked earlier by someone else," and "It looks like a Y, so it probably is a Y." At an organisational level, ETTO rules include negative reporting (where the absence of reporting is interpreted as indicating that everything is okay), cost reduction imperatives (which increase efficiency at the cost of thoroughness), and double binds (where the explicit policy is "safety first" but the implicit policy is "production takes precedence when goal conflicts arise").

As a recent example of this, U.S. Secretary of Transportation Elaine L. Chao, on July 18, 2019, commented on how the Federal Aviation Administration was approving Boeing's 737 MAX for flight again. She pointed out that the FAA "is following a thorough process, rather than a prescribed timeline . . . the FAA will lift the aircraft's prohibition order when it is deemed safe to do so." This is a clear, if rare, example of thoroughness taking precedence over efficiency.

A special case of the ETTO principle, which is practically a phenomenon on its own, is the *confirmation bias*. This describes the tendency of people to search for, interpret, favour, and recall information that confirms their pre-existing beliefs, hypotheses, or assumptions. The confirmation bias affects how people gather or remember information. It clearly contributes to efficiency because the search for evidence or "proof" stops when an example is found that confirms expectations. It can likewise be seen as an illustration of the filter response to IIO. This clashes with the philosophical falsification principle, which proposes that for something to be scientific it must be falsifiable. In other words, when trying to ascertain that something is the case, we should also look for negative examples, for something that goes against the underlying assumptions. Doing so usually involves considerably more effort and is therefore avoided unless absolutely necessary. Confirmation biases contribute to overconfidence in personal beliefs and can maintain or strengthen beliefs in the face of contrary evidence. Poor decisions that result from these biases are unfortunately all too easy to find, not least in political and organisational contexts.

The confirmation bias comes in two varieties: factual and hypothetical. In the factual confirmation bias, people will search for facts that support their beliefs or assumptions. Extreme cases are climate sceptics and flat earthers. The factual confirmation bias is also often found in political arguments but is, of course, a short circuit of logical thinking. In the hypothetical confirmation bias, people will look for positive outcomes of their intended actions or plans and use that as "evidence" that their plans will work. This is unfortunately something that is used frequently in planning or organisational changes of one kind or the other, such as the introduction of lean management or new campaigns for productivity or well-being at work.

Consequences of psychologically based fragmentation

The previous sections have described some of the limitations that affect how people think and reason about the world they live in and how these limitations may have a biological, cognitive, or social basis. To compensate for and partly overcome limitations, people have developed a number of heuristics that generally make it possible to "muddle through." Several of these have already been mentioned – responses to IIO, bounded rationality, efficiency-thoroughness trade-off, decision heuristics – and more can easily be found in the rather rich literature on this topic. They can also all easily be recognised in how individuals perform, how people act collectively, and how organisations perform. The heuristics contribute to making life more manageable and making it possible to cope with complexity, but they also contribute to and often increase the fragmentation, which thereby can become a hindrance for the effective management of change.

Most of these heuristics are so common that they are accepted as normal practice and are therefore no longer noticed. In some cases they are even attributed to the reality that provides the context for work and actions rather than to the shortcomings of how humans think and reason. They are seen as having an objective existence rather than as being a consequence of how thinking and reasoning have become adjusted to the conditions. If the underlying limitations have a biological basis it is difficult or even impossible to do something about them. But when they exist in the mind, when they have a cognitive basis, and, in particular, when they have a social basis, it may be possible to do something about them – provided, of course, that their existence is accepted and acknowledged. The following sections will present and describe the most important limitations that have a social basis.

Dichotomous or binary thinking

William James, who was mentioned earlier, provided the following description of how humans experience the world:

> [A]ny number of impressions, from any number of sensory sources, falling simul-
> taneously on a mind WHICH HAS NOT YET EXPERIENCED THEM SEPA-
> RATELY, will fuse into a single undivided object for that mind. The law is that all
> things fuse that can fuse, and nothing separates except what must.
>
> (James, 1890, p. 488)

A baby has no experience and no way of discriminating the information that assails it, and all the sensory impressions therefore end as a "great, blooming, buzzing confusion" (Ibid.). As humans grow up they become able to avoid that because they learn to recognise a variety of patterns or constructions (James's term) that bring some sort of order to the data chaos (somewhat like a combination of the IIO strategy of reducing precision and cutting categories in Table 3.1). The constructions are the habitual ways of describing, and hence of seeing or perceiving, the surrounding world. They serve the individual because they introduce some kind of simplification – and therefore also fragmentation – just as they serve an organisation or collective because they provide a basis for a shared (intersubjective) understanding and communication. The constructions in particular allow people to address the problems one by one as if they were independent or existed independently of each other. Because this is done so often, the fact that they are not solvable in that way is gradually forgotten. The continued use of these constructions makes it difficult to understand that things are interconnected and that attempts to manage them without taking the interconnections into account will only make this worse and will only create new problems.

Following James's argument, humans tend, or prefer, to see the world as relatively undivided and, at most, as one concept together with its contradiction or opposite. Unless a construction is everything – literally everything – there must logically be something that is not a part of it. There is therefore always something that is in contrast to or the opposite of the construction. This is known as dichotomous (from Greek *dikhotomia*) or binary (from Latin *bini* or *binarius*) thinking. Today's language and communication are full of dichotomies – "us versus them," "black versus white," "friend versus foe," "safe versus unsafe," etc. This has even given rise to social identity theory (Tajfel & Turner, 1986), which states that people tend to divide the world into the in-group (us) and the out-group (them), and the in-group will discriminate against the out-group to enhance their self-image.

Dichotomous or binary thinking is easy on the mind and therefore seems to be helpful in overcoming the limitations of attention, etc. It is easy for humans to focus on a single concept or single idea, and it is also easy to apply a single principle of reasoning – notably linear causation. If the phenomenon is simple in itself, then a simple explanation is apparently justified; simple problems *may* have simple solutions. But if the phenomenon is not simple, whether it is called complex, complicated, intractable, or non-trivial, then a simple explanation is impossible. Attempts to provide simple explanations for complex problems unfortunately do not make the problems simple; they only result in solutions that are incapable of actually improving anything. Despite that, humans like and prefer simple or *monolithic* explanations because they fit well with the various limitations in thinking and reasoning. This is easiest to see in the case of causality and the search for root causes. Other prevalent examples of monolithic thinking are shown in Table 3.2.

Table 3.2 can also be described as showing contrafactual causes, in analogy to contrafactual conditionals. The reasoning according to contrafactual causes is that "if X had been present, then Y would not have happened." In other words, the cause is a lack of something that conveniently is "invented" as the need arises. Contrafactual causes are the standard explanations where the cause is seen to be "a lack of something" – lack of trust, lack of situation awareness, lack of communication, lack of safety culture, and so on.

It would indeed be nice if the problems of managing organisations, whether with regard to productivity, quality, safety, reliability, or anything else, could be treated and solved issue by issue. We can call this for the simple problem – simple solution fallacy – though it may not have been a fallacy in former times (cf. the account of the history in Chapter 2). In the 1930s, organisations and production systems could be considered on their own, and the dominant problems could be solved on their own. Systems were not tightly integrated and were not tightly coupled. But today they are.

The problems of today are often explained by pointing out that systems have become complex and that they therefore are difficult to manage and control. This is easily recognisable as yet another monolithic explanation. But the problem is not complexity as such (whatever that may be), it is rather that the psychologically based fragmentation makes it impossible to build an adequate understanding of systems. The means – concepts and

Table 3.2 Monolithic causes and monolithic solutions

Monolithic causes (root causes)	*Monolithic solutions (silver bullets)*
Technical failures	Design, construction, maintenance
Human error	Train, automate, redesign, simplify
(Lack of) safety culture	Improved safety culture
Deviations from norms	Compliance
Brittleness	Resilience
. . .	. . .

theories – that are available to describe systems and work tend to be oversimplified and therefore inadequate for today's problems. That this is so can hardly be a surprise to anyone.

Humans have a preference for thinking of the world in terms of single themes or issues. Examples can easily be found in politics and science – and in business as well. We tend to describe the world as black or white and to use that as the basis of communication. This may be due to the mistaken idea that it is easier to communicate something that is binary than something that is not, or that it is easier to control something that is binary than something that is not – i.e., whether something has passed a threshold.

Linear causality

These limitations in thinking make it natural to reason in a linear way. This means that reasoning about a given phenomenon may be either a forward extrapolation of the consequences in a step-by-step fashion or a backwards analysis of causes in a step-by-step fashion – and usually only a few steps in either direction. It is necessary to think of the consequences when planning to do something and to imagine what the outcomes could be as well as possible. Here the cause is represented by the planned action or intervention and the effect by the intended or expected outcomes. Humans like to envisage the desired outcome (confirmation bias) and tend to disregard the undesired (side effects). They also optimistically assume that the intended action is the only or main determinant of what will happen next (treated in more detail in Chapter 4). They do so despite knowing from everyday experience that the future is not linear or predictable. Neither is it a secret that if plans look too far ahead the number of alternative developments that have to be considered easily exceeds what the mind can cope with. This is probably why wishful thinking often takes over – for instance by using a safety case as a way to argue that a system is acceptably safe for a specific application in a specific operating environment. The same happens when trying to understand why something has happened, i.e., when thinking or reasoning backwards. Already Aristotle argued that "we think we do not have knowledge of a thing until we have grasped its why, that is to say, its cause" (Physics, 194b 17–20). Here the assumptions of cause and effect are even stronger. When something has happened, there must have been something that happened before it that therefore was the cause.

In both cases people are led astray by the fact that events are always described in relation to time using a timeline. Because time is one-dimensional, the position of two events next to each other on a timeline is deceptive. It looks as if the preceding event was the cause and that the following event is the effect. In combination with binary thinking this logically leads to the idea of the true cause, which Aristotle called the efficient cause. Jumping ahead in time, we see that reinstated in Heinrich's notion of the chain of causes illustrated by the domino model. Tolstoy had the following to say about the limitations of the human mind:

> Man's mind cannot grasp the causes of events in their completeness, but the desire to find those causes is implanted in man's soul. And without considering the multiplicity and complexity of the conditions any one of which taken separately may seem to be the cause, he snatches at the first approximation to a cause that seems to him intelligible and says: "This is the cause!"
>
> (Tolstoy, 1993; org. 1869, p. 777)

Shewhart seems to have agreed:

> As human beings, we want a cause for everything but nothing is more elusive than this thing we call a cause. Every cause has its cause and so on *ad infinitum*. We never get quite to the *infinitum*.
>
> (Shewhart, 1931, p. 131)

Synesis – Breadth–Before–Depth

Historically based fragmentation and psychologically based fragmentation both favour a Depth-Before-Breadth (DBB) approach to problem-solving and management. Historically based fragmentation leads to a focus on a single issue and the pursuit of that in isolation, rather than a broader understanding of the dependencies that characterise both internal and external conditions. Psychologically based fragmentation limits the number of things – aspects, criteria, alternatives – that can be considered at any one time and also favours a linear cause-effect type of reasoning. One consequence in both cases is a set of models and methods that are too simple in the sense that there is a significant discrepancy between what they can capture and cope with and what actually happens. This is the same kind of discrepancy that exists between Work-as-Imagined and Work-as-Done. Despite their psychological convenience and appeal, simple models and methods do not provide adequate support for the management of organisations that are even partly intractable. The only sensible solution is therefore to develop better models and methods – in other words, to develop or build descriptions that better correspond to or match the "real" world. This is also in agreement with the law of requisite variety, which states that the variety of outcomes (of a system) can only be decreased by increasing the variety in the controller of that system. The law of requisite variety will be described in more detail in Chapter 4.

Problems are solved in a Depth-Before-Breadth manner when a chosen line of reasoning or analysis is pursued to the end before alternatives are considered. A Depth-Before-Breadth strategy narrows the search and is therefore easier to use in terms of mental effort, but it also reinforces fragmentation. The alternative is a Breadth-Before-Depth (BBD) approach, where several things or alternatives are considered at the same time. BBD requires the ability to step back and consider the problem as a whole, to look at it from several different angles, before a detailed analysis is begun. A BBD strategy makes it more likely that a solution is found and less likely that efforts are wasted by going in the wrong direction. It favours thoroughness over efficiency, which is always an advantage in the long run. It specifically helps to overcome the fragmentation that dominates current approaches to organisation management.

The principle of synesis, the principle of unifying the different issues rather than separating them, is clearly consistent with a Breadth-Before-Depth approach. Synesis makes clear that it is necessary to abandon the search for simple solutions and give up hope that there is a silver bullet that will magically solve every problem no matter what it is. But synesis is not in itself a solution. If it were, it would violate the principle it tries to promote. Synesis is, rather, an indication of how a solution to the management of intractable organisations can be developed. Synesis is about how to improve productivity and efficiency, but not just about how to reduce waste. It is about how to improve quality, but not just about how to reduce defects and deficiencies. It is about how to improve safety, but not just about how to reduce accidents and adverse outcomes. And it is about how to improve reliability (availability), but not just about how to reduce vulnerability or increase linear redundancy. Altogether synesis it is about how to improve something, but not about how to increase something. *Increase* implies looking at something by itself, but *improve* implies – I hope – that issues and concerns become unified and are looked at together, as a whole. The striving to achieve unification is a hermeneutical process that opens up new interpretations, and the process of unifying can therefore in some sense be more important than the result.

4 Fundamentals of change management

It is the nature of all things to change, to perish and be transformed, so that in succession different things can come to be.

Marcus Aurelius, Meditations, Book XII:21

Introduction

To achieve a goal or an objective, either to improve something or to prevent something from happening, it is necessary to ensure that the state of the system changes from whatever it is at present to become whatever it should be in the future. To do so requires the ability to manage how changes take place, which essentially means the ability to control them. This is abundantly obvious from the dictionary definitions of the verb *manage*, which include "be in charge of," "run," "head," "direct," "control," "lead," "govern," "rule," "command," "take forward," "guide," and "be at the helm of," and so on and so forth. The essence of managing something is the ability to make sure it develops as intended, as foreseen, or as planned.

Among the multiple definitions just mentioned, the best characterisation of what it means to manage something is possibly to "be at the helm of." This is a maritime expression for being able to control the movement – direction and speed – of a vessel. For this to be possible, three conditions must be fulfilled. It is first of all necessary to know what the destination or target position is. Without that it is impossible to set the course towards the target and impossible to know when the target position has been reached. Second, it is necessary to know what the current position is and to keep track of how it changes over time. Without knowing the current position it is impossible to know whether the distance to the target is being reduced, just as it is impossible to plan and prepare the continued voyage. Finally, it is necessary to know how to actually control the vessel and how to ensure that the movement is in the right direction and that the speed, or rate of change, is what it should be.

The voyage metaphor

Applied to the actual control of physical movement, whether of a vessel, car, aeroplane, etc., it is clear that the three conditions are required as illustrated in Figure 4.1. Consider, for instance, being at the tiller of a small sailboat or yacht. (Readers who have never tried to do that can instead think of driving a car in an unfamiliar environment – for instance during a holiday in another country.) Here it is clearly important to know where you

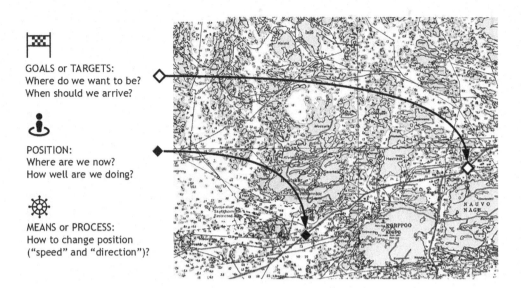

GOALS or TARGETS:
Where do we want to be?
When should we arrive?

POSITION:
Where are we now?
How well are we doing?

MEANS or PROCESS:
How to change position
("speed" and "direction")?

Figure 4.1 The voyage metaphor.

want to arrive, either as a waypoint or as a specific location. It is equally important to know where you are and whether the journey is going as planned so that you will arrive in time at wherever you want to be. Today this is frequently done using GPS, but not so long ago navigation relied on physical sea charts or, in the case of driving, on roadmaps. Finally, if you are at the tiller of a sailboat or at the wheel of a car, it is important that you know how to control the vessel or the vehicle, how to change the heading or direction, and how to change the speed – although for a sailboat the degree of control over course and speed can be somewhat limited. For a car it is easier and more direct.

The importance of knowing the position

The importance of knowing exactly where you are is dramatically illustrated by two naval accidents. On October 22, 1707, a British Royal Navy fleet lost four warships in severe weather off the Isles of Scilly. Between 1,400 and 2,000 sailors lost their lives aboard the wrecked vessels. The main cause of the disaster has been attributed to the navigators' inability to accurately determine their longitude, which meant that they had to rely on dead reckoning. The disaster was one of the reasons for the Longitude Act in 1714, which established the Board of Longitude and offered large financial rewards for anyone who could find a method of determining longitude accurately at sea.

The Scilly naval disaster happened long before GPS was imaginable. But even after GPS became standard equipment on ships, the grounding of the cruise ship Royal Majesty showed how important it is to be aware of your position. The Royal Majesty sailed from Bermuda bound for Boston on June 9, 1995. Shortly after departure, the cable from the GPS antenna became disconnected so that the GPS receiver no longer received satellite signals. The GPS receiver automatically switched to dead reckoning mode where the autopilot tracked the GPS "data" until the ship grounded on the Nantucket shoals some

34 hours later, 17 miles west of course. No one was injured by the grounding, but the cost in terms of lost revenue was substantial.

The loss of signal was not clearly announced when it happened and therefore went unnoticed by the crew. But neither did the crew cross-check navigational data with other independent sources, such as the LORAN-C navigation or celestial, radar, or compass bearings, even though it was required by the procedures. The second officer was actually so sure of his track position that he ignored several clues that suggested otherwise. This can be seen as another example of the confirmation bias discussed in Chapter 3. Here, as in many other cases, people created explanations that maintained their conviction because it was easier and more convenient than changing it. Yet dead reckoning is no substitute for actually knowing the position.

It is, of course, important to know one's position in all cases regardless of whether the travel is physical or symbolic. A business may founder as easily as a ship if those in charge do not know precisely where they are. Managing a company, a business, or even a country is likely to fail if your assessment of the position is wrong. The most dramatic example of this in recent times is perhaps David Cameron's failure to correctly assess the "position" of the British voters when the Brexit referendum was held on June 23, 2016. Less dramatic examples are easy to find. To name just one, the American market research company Forrester in August 2019 produced a report with the following title: *Why Brands Must Bridge The Gap Between What They Think They Know And What Customers Really Want.*

On roadmaps

The voyage metaphor is indeed often used to describe changes in organisations and how management can develop a "roadmap" to take them to their destination. Expressions such as "the CEO in the driver's seat" are also common. But while the analogy with a roadmap sounds attractive, a voyage by sea is actually a better reference than a voyage by land. A roadmap literally shows the roads and topography of a region or country. It shows you where you can drive and can be used to decide which roads should be taken to get from the current position to the destination. In the context of change management, the term is used to describe a strategic plan that defines a goal as well as the major steps or milestones needed to reach it. If the comparison is a travel by land, then it is reasonable to assume that the roads exist and that they are there as long as the journey lasts. There may, of course, be roadworks or other hindrances on the road, but in such cases the roadmap can be used to find a detour that will ensure that the voyage ends successfully even though it may take longer than originally planned.

Change management is, however, more like travel by sea than by land. For a sea voyage there may be waypoints, markers, or buoys, but there are no physical lanes or roads. Where the land usually remains stable – flash floods, snowstorms, and earthquakes excepted – the sea is never stable. There may be strong wind or storms, currents, waves (even freak waves), hidden reefs or shoals, etc. And if the vessel is driven by the wind rather than by an engine, the wind may be against you or there may be lulls, which means that the planned route must be abandoned. For a voyage by land the topography is known and permanent – at least during the time it takes to complete the travel. For a voyage by sea no such permanence or stability exists. So for a journey with an organisation, whether it is change management, the safety culture journey, the learning organisation journey, the journey to organisational excellence, etc., we should think of it as a voyage by sea rather than by land and acknowledge that there are no maps to help us and, furthermore, that we have to set the course ourselves. Neither are the three conditions – the target, the position, and knowing how to make a change – as easy to meet as in making a physical journey.

Change management as a voyage

When the change refers to something that can actually be measured in one way or another, such as the number of units produced, it makes obvious sense to talk and think about the target as well as the current position or state. For the production of something concrete, whether in the form of matter or energy, the process is also usually known so well that it is possible to determine how a change can be made. As long as the outcomes are concrete and measurable, the three conditions are easy to connect to practice.

The situation is considerably different in the case of changes where the output is in terms of a quality or a notional state or condition and hence non-material or intangible. The four issues that were introduced in Chapter 2 and are used throughout this book – productivity, quality, safety, and reliability – can be used to illustrate this. Every organisation is concerned with at least two of them – typically productivity and quality or productivity and safety – and many need to consider the other issues as well. Productivity, for instance, is necessary to generate revenue so that an organisation can survive financially. Quality is likewise necessary to maintain customers and possibly even to add new ones as well as to limit costs for repairs and replacements. Safety is necessary because accidents and incidents can harm workers and/or customers, because adverse events can interrupt or disrupt production for shorter or longer periods of time, and because they may lead to a bad reputation and hence a loss of business. Finally, reliability is necessary to ensure an acceptable level of performance or functionality both now and throughout the lifetime of the product.

In terms of the three conditions, it is clearly necessary to know what the target is. This is needed to decide which changes are necessary and to provide a point of reference during the journey. On a general level there can be production targets, safety targets, quality targets, reliability targets, and so on, but each of these must usually be specified further to be operational. In some cases targets may sound very concrete – like a 12% increase in productivity, 2% reduction in annual cost, or zero accidents, for instance – but such precision can be illusory.

The second condition is to know what the current position or state is, or how well an organisation is doing relative to a specific target. Repeated measurements of the position are necessary to ensure that the change goes in the intended direction and that the rate of change is as expected, although such expectations often represent the triumph of hope over experience, to paraphrase Samuel Johnson. But determining the position with regard to targets such as safety, reliability, learning, or excellence can be difficult because valid measurements, and even good performance indicators, are hard to find. Measurements are also usually lagging and sometimes have a considerable delay. Additionally, due to the costs and efforts required to make the measurements, the frequency may be inappropriate for the needs of managing the process.

The third condition presents an even more serious problem; namely, how an organisation can be controlled. How can the notion of the "direction of change" be translated into something practical or tangible and how can it be realised? Similarly, how can the notion of the "rate of change" be made operational and how can it be realised? There are no obvious means of control for an organisation, either generally or particularly. Management dashboards unfortunately do not include the appropriate means of control. In practice, interventions are either derived from legacy, replicating general industry practice, or imitating current trends. There is furthermore a serious lack of good models of organisations. There is, of course, no shortage of impressive-looking diagrams or charts that describe organisations in terms of structures, communication channels, roles, norms, interactions, and so on. Sociograms and social networks may go some way towards actually depicting what goes on, although not to the extent that they can serve as reliable descriptions of how an organisation functions. The sad truth is that the most realistic model of an organisation probably is the black box, i.e., a system that is described in terms of its inputs and outputs and their relations but without any knowledge of its internal workings. The reason for this is that organisations are never designed or built in the way we can design and build a technological system such as a cruise liner, a cement factory, or the Large Hadron Collider. There will in many cases be an intention or a general plan or blueprint for an organisation – or more commonly an already existing organisation whose performance is emulated or simulated – but in real life organisations grow in ways that are far from being completely understood. (One fundamental reason for this is that conditions and roles are always underspecified and hence completed by clever adjustments. Another is that organisational life is governed by Parkinson's laws.)

Yet in order to control a process – and managing organisational changes of productivity, quality, safety, etc., *is* controlling a process – it is necessary to have a good understanding of what actually goes on. In other words, it is necessary to know how an organisation works or functions in a concrete rather than abstract sense. More will be said about this in Chapter 5.

Steady-state management versus change management

Organisations large and small generally fare better during periods of stability than during periods of turbulence simply because stability is a precondition for perfect predictability. If the surroundings are stable (economic, political, supply and demand, technological progress, etc.), the role of management can in many cases be reduced to ensuring that an organisation continues to function, that it meets its targets or standards for performance, as effectively as possible. This is called steady-state management. (A system or process is said to be in a steady state if the variables that are used to characterise and define the behaviour of the system or process do not change over time. Current system behaviour can therefore be expected to continue into the future.) Steady-state management is about continuing to do what we do to maintain the current target. This is completely different from trying to attain a new target, trying to do something differently, or doing it differently.

Even when the target does not change, it is necessary to manage an organisation's performance because there always will be some kind of internal or external variability and hence reduced predictability. Steady-state management need, however, focus only on compensating for perturbations or variabilities that are seen as being unacceptably large, as in the way that quality management concentrates on the assignable causes rather than the chance causes (cf. Chapter 2). This was feasible when the production of something or the provision of a service took place under stable conditions a century or so ago. But the purpose of management today is rarely limited to ensuring stable production or to maintaining constant performance and quality under stable conditions. The purpose is more likely to ensure stable production and performance under unstable and partly unpredictable conditions and hence to ensure that an organisation is able to meet current as well as future challenges – threats as well as opportunities. Management should be prepared to face a never-ending series of changes in order to cope with unstable conditions and uncertainty arising, intentionally or unintentionally, from competitors, new technologies, regulations, services, and customer demands.

Even if the objective is to maintain the *status quo* rather than to do something in a different way or to do something differently, it is still necessary to manage change. The reason is simply that no organisation and no organisational environment are completely static and unchanging. If something really never changed it would, of course, not be necessary to manage it. Just letting it remain as it was would be sufficient. This is typically the case for physical objects or artefacts if we only consider them for relatively short periods or durations of time. I do not, for instance, need to manage my dining room table because it does not change from one day to the next – and it may not even change from one year to the next. But seen over a sufficiently long period of time it will change and gradually deteriorate. It nevertheless does not matter much to me because it will take longer than my expected remaining life span; I can therefore, for all intents and purposes, consider it stable. By contrast, I do need to manage the flowers in the vase on the table since, without watering, they will wither and die. They will still do so eventually, so I can only maintain them for a limited period of time.

If instead of looking at dining room tables and flowers we look at organisations or the machinery or technology means by which organisations carry out their functions, we can see that these clearly change even if nothing is done. Material items – machines, supplies, reserves, etc. – deteriorate if left to themselves, which is why they need to be maintained or replenished. Organisations and the people in them change, not because they deteriorate but because they, by their very nature as social systems, are unstable – because they learn and because they forget. This is why safety or quality campaigns must be repeated from time to time. Organisations and the people in them also change due to influences from their surroundings, primarily because the surroundings include systems and organisations where intentional changes are made – sometimes competitive, sometimes adversarial, sometimes supportive. The surroundings may, of course, also change for other reasons, such as thermodynamic or anthropogenic entropy (cf. later discussion). Change management is therefore always required (1) to eliminate or counteract unintended changes or variability (cf. classical quality control [maintaining the steady or stable state]), (2) to change the way an organisation works so that performance meets a new target or intention, or (3) to cope with an uncertain future – threats as well as opportunities.

Thermodynamic entropy

The purpose of controlling a system can be seen as an attempt to reduce entropy, which means to reduce disorder and lack of predictability. Entropy, generally speaking, is an expression of the degree of disorder or uncertainty in a system. In classical physics *entropy* is defined as a thermodynamic quantity that represents the unavailability of a system's thermal energy for conversion into mechanical work. The less technical interpretation is that entropy is the degree of disorder or randomness in the system. Clearly, the more disorderly a system is, the fewer resources will be available for useful activities. The second law of thermodynamics states that the entropy of a system that is not subject to external influences, a so-called closed or isolated system, never decreases. Every type of system – physical or social, technological or biological – must constantly battle against forces of various kinds, internal and external, that reduce stability and constancy. Waste can be seen as entropy, as in Lean. Lack of quality can be seen as entropy, as in statistical quality control or Kaizen. Accidents can be seen as entropy, as in Safety-I. In other words, the world gets more and more disorderly.

In systems theory a system is defined as *closed* if there is no exchange of inputs or outputs across the boundary that separates the system from the surroundings. If a closed system is left to itself, the degree of disorder or uncertainty will gradually but inevitably increase and never decrease. The systems we manage – the organisations, businesses, and services we rely on – are obviously open rather than closed because there is an exchange of inputs and outputs across the boundary. We use this exchange to manage the system and to keep the increasing entropy temporarily at bay. Systems are built or established with a specific purpose in mind, and we actively do our best to manage them by providing inputs – data, guidance, instructions, directions, etc. – and by evaluating, measuring, and assessing the outcomes to make sure that the system performs as intended. There are therefore, in reality, not just one but two systems: the system that is being controlled or managed (called the *target system*) and the system that controls or manages it (called the *controller*). The target system, which we can call $system_1$, is by definition an open system because the controller provides inputs to it and receives outputs from it. The target system and the controller together can be seen as constituting a system in their own right, which we can call $system_2$. This obviously raises the question of whether $system_2$ is closed. If it is open and therefore controlled, then $system_2$ and its controller together constitute another system, which we can call $system_3$. This way of nesting systems can be continued for a long time, but not forever. Sooner or later we will, in practice if not in theory, reach a point where the system under consideration, called $system_n$, is closed or isolated in the sense that it is left to itself with nothing to control it. In other words, while we may establish and maintain sets of systems that are controlled, it will not be possible to do so *ad infinitum* (cf. Figure 4.2).

Examples are easy to find – the world economy, the environment (global warming), the population of Earth, pollution, etc. $System_n$, however large n is, will be closed and

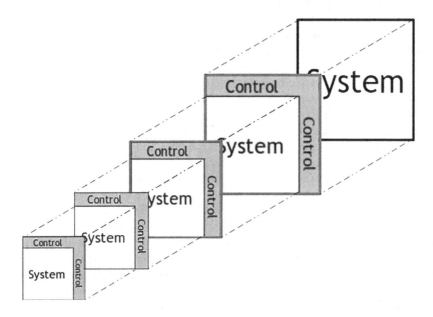

Figure 4.2 Nested systems and their controls.

hence subject to the second law. Since system$_n$ comprises a system$_{n-1}$ plus its controller, the entropy will affect not only system$_{n-1}$ (the target system) but also its controller, thereby contributing to the overall degradation of performance. The striving locally to reduce entropy, the striving to be in control, the striving to ensure that an organisation performs as intended and produces the desired outcomes as far as possible is at the heart of synesis.

In practice we can manage most of the systems we in one way or another have introduced and therefore keep entropy reasonably well under control. The entropy, the disorder or the lack of predictability, will, however, always and irrevocably increase, which means that the demands to control also will increase.

Anthropogenic entropy

The classical interpretation of entropy is used in many fields, such as information theory, computing, cosmology, and economics. But for socio-technical systems there is another and notably more important form of entropy that we may call *anthropogenic entropy*. This refers to the increasing disorder and decreasing predictability that are due to our fragmented understanding of the world and hence the inadequate steps that are taken to restore order and the inaccuracy of the interventions that are made. In the case of socio-technical systems, the surroundings – the external forces – are themselves socio-technical systems. These are never stable or constant but always develop and change for better or worse. Anthropogenic entropy to a large extent exists because of the psychological reasons for fragmentation. Because we can never completely comprehend the whole situation, and because of the limitations in how far we can look ahead and how thorough we can be in considering alternatives, there is a need for constant adjustments and corrections that, because they are always approximate, become a source of entropy. Examples are (too) easily found in politics (national and international), trade and commerce (again, national and international), interpersonal relations, etc.

Anthropogenic entropy thus comes about when we, as humans, try to establish order but fail to do so. It is a condition for survival that there are not too many surprises, and the way to ensure that is to seek or create some kind of order or stability. In the beginning, when we were hunter-gatherers and nomads and the only inanimate agents that were necessary to consider were the animals – those we hunted and those that hunted us – the world was predictable because it was relatively stable. Conditions worsened as societies and civilisations started to develop since it was now also necessary to consider other human beings. This may have happened during the Axial Age, the Bronze Age, or even earlier. The development involved animal husbandry as well as the division of labour, through which people acquired specialised capabilities and either formed coalitions or traded to take advantage of the capabilities of others in addition to their own. Humans have the dubious quality of being able to partly outguess those who try to control them and to use that as part of their own attempts of control. Due to the effects of psychological fragmentation, both trying to control others and outguessing others will be imperfect, which increases the entropy. In the beginning the level of anthropogenic entropy was low and increased slowly. But as civilisation progressed the entropy increased rapidly due to several deviation-amplifying mutual causal processes (Figure 4.3), a.k.a. positive feedback loops (Maruyama, 1963) (the basic notation has its roots in path analysis and directed graphs but was much enriched by Maruyama's description of what he called the second cybernetics). The same type of thinking is found in fields such as system dynamics (Forrester, 1971) as well as causal loop diagrams, with a more extensive discussion provided in

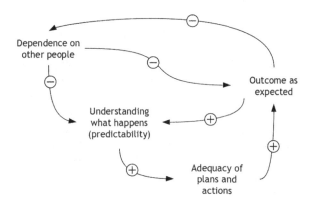

Figure 4.3 A causal diagram of anthropogenic entropy.

Hollnagel (2012). Causal loop diagrams made it possible to show the two ways in which processes could influence each other. The first was called deviation-counteracting and corresponded to the classical negative feedback loop. The second was called deviation-amplifying and corresponded to a positive feedback loop. A classic example is a bank run that happens when many clients withdraw their money from a bank because they believe the bank may cease to function in the near future. Because more people withdraw cash, the likelihood of default increases, which triggers further withdrawals, which increases the likelihood of a default, and so on.

In Figure 4.3 the arrows show that there is a relation between two states or conditions, and the "+" and "-" signs represent the nature of the relation. A "+" sign between A and B means that the two states are directly proportional: if A increases then B also increases, and if A decreases then B also decreases. Similarly, a "-" sign between A and B means that the two states are inversely proportional. Thus, as the understanding of what happens is improved, the adequacy of plans and actions will also improve and the likelihood of unanticipated outcomes will decrease. If the number of unanticipated outcomes increases, then anthropogenic entropy increases and the understanding of what happens decreases. Because entropy and anthropogenic entropy always will increase, this kind of causal loop diagram shows that our attempts to control the world and make it predictable in the long run are doomed to fail.

Law of unanticipated consequences

One of the most conspicuous manifestations of anthropogenic entropy is the occurrence of unintended and unexpected outcomes, more commonly known as the problem of unanticipated consequences. If the outcomes of controlled interventions, of changes or corrections, do not lead to the expected outcomes, that clearly affects the orderliness of the system in a negative way that disrupts, or at least impacts on, future actions.

The occurrence of unanticipated consequences, the fact that situations and events do not always develop as we had intended or hoped for, is undoubtedly as old as humankind itself. ("Accidents," as Mr. Micawber told David Copperfield, "will occur in the best-regulated families; . . . they may be expected with confidence, and must be borne with philosophy.") In modern times, the phenomenon came to prominence with the publication of a delightful paper entitled *The Unanticipated Consequences of Social Action* (Merton, 1936). This presented a cogent analysis of the phenomenon and identified the following factors or conditions that seem to contribute to the occurrence of unanticipated consequences:

- Ignorance: Inadequate or insufficient knowledge about the possible consequences of actions and decisions. A particular form of inadequate knowledge relates to the understanding of how the system functions, how an organisation works. Even a century ago, there was a discrepancy between organisational models and the actual organisations. This discrepancy has only grown as society and organisations/systems have continued to develop and become more complex.
- Error: False reasoning; limited scope (negligence, bias). These are essentially the issues that have been described in Chapter 3 as part of the psychological causes of fragmentation.
- Imperious immediacy of interest: A dominant or overriding interest in primary/immediate results that leads to a disregard or even denial of side effects. Promises by politicians during election campaigns – or even in general – are good examples of this.
- Basic values: The importance of sustaining basic values (criteria) means that long-term consequences are neglected. In these cases it is felt necessary to take certain actions to uphold fundamental values.
- Self-defeating predictions: Predictions become a new element of the situation, thus tending to change the initial course of developments.

Taken together, the law of entropy and the law of unanticipated consequences mean that the number of unanticipated consequences will always increase (cf. Figure 4.3). The more disorderly and uncertain the world and the systems in it become, the more likely it is that actions and changes will have unanticipated consequences. This will create a vicious circle – or a "whack-a-mole" condition – where the pressure to increase efficiency, to get things right, to improve quality, or to quickly to recover from accidents competes with and dominates the need of the thoroughness required to get a reasonably correct understanding of how everything works. As predictions become less precise, the number of unanticipated consequences or "errors" will increase, making it even more urgent that control is regained. A growing number of unexpected outcomes increases the demands on how the system is managed and hence the demands on the controller. This relationship has been described by yet another principle known as the law of requisite variety.

The law of requisite variety

It is hardly necessary to argue that it is only possible to manage a system if we understand how it works or functions. To invoke the voyage metaphor again, the purpose of management is to ensure that we move closer to the target in an orderly manner, which, in the terms of the voyage metaphor, means that we know where we want to be, we know where we are, and we can control the speed and direction of the change. Controlling the "movement" of an organisation can only be done if we understand how an organisation works and what its "inner mechanisms" are. Insufficient knowledge about that represents a condition of ignorance – usually partial but in rare cases almost complete – that inevitably will lead to the occurrence of unanticipated outcomes and hence a gradual loss of control.

The law of requisite variety (LoRV) was formulated in cybernetics in the 1940s and 1950s (Ashby, 1957) and was lucidly explained in a paper entitled *Every Good Regulator of a System Must Be a Model of That System* (Conant & Ashby, 1970). The law is concerned with the problem of regulation or control and expresses the principle that the variety of a controller should match the variety of the system to be controlled, where the latter is usually described in terms of the variety of a process plus a source of noise or disturbance. The implication of the law is that the number of things that can happen, the degrees of freedom of the output from a system, must be matched by the number of states or conditions the controller is able to recognise and respond to. In simple terms it means that if something can happen that the controller, or management, did not expect or was not prepared for, then control may be lost. This is not an uncommon situation, as illustrated by the many times people have been surprised, whether by an unexpected hindrance or development, an "unusual" accident, or a transformative political event such as Brexit. The LoRV proposes that the variety of the outcomes (of a system) can only be decreased by increasing the variety in the controller of that system. Effective management is therefore not possible if the managing system, the controller, has less variety than the system, hence the title of the Conant and Ashby paper just mentioned.

Another way of expressing the law of requisite variety is the famous advice found in Sun Tzu's *The Art of War*: "knowing your enemy and yourself will help you to win; not knowing your enemy but knowing yourself means you are uncertain; knowing neither your enemy or [*sic*] yourself means you are sure to lose." In other words, knowing what your enemy may do or is capable of doing is necessary to prepare appropriate responses. In politics it is similarly necessary to know what your opponent may do, just as in business it is necessary to know what your competitor – or even your business partner – may do.

Returning to the voyage metaphor, the law of requisite variety means that it is impossible to control or manage changes in an organisation unless the understanding of how an organisation functions is so thorough – or extensive – that there will be few, if any, surprises. In other words, it is necessary that an organisation is understood so well and that the management of it is so precise and careful that unanticipated consequences are unlikely to occur. It may have been obvious to the cyberneticians sixty years ago that the model or understanding should be of how the system functions: cybernetics is defined as the scientific study of control and communication in the animal and the machine, and the focus of the LoRV is, after all, on variety or variability. Yet models of organisations are today usually expressed in terms of their structure or architecture, with the unspoken assumption that there is a simple, or even a 1:1, relation between structures and functions. Each part (element or component) of the structure and each part or unit of an organisation has a well-defined function – similar to the relation between parts and functions in a mechanical system. That this rarely, if ever, is the case for organisations understood as socio-technical systems is recognised by the crucial distinction between the formal and the informal organisation. The formal organisation refers to how the organisation is imagined to function according to clearly laid out rules, goals, division of labour, and clearly defined hierarchy of power. The informal organisation refers to the interlocking and often implicit roles and relationships that govern how people work together in practice. Organisational models blissfully ignore that it often may be difficult, and sometimes even impossible, to identify or recognise distinct "structures." Yet it is impossible to regulate or manage something without a model or an understanding of how that something functions. To paraphrase the distinction between Work-as-Imagined and Work-as-Done, we need a similar distinction between the Organisation-as-Imagined and the Organisation-as-Existing – the actual organisation. The model of the actual organisation must clearly comprise both individual and organisational functions at all levels and the ways in which they are coupled or interdependent.

Challenges from a shrinking world

For most of the time that civilisations, societies, and organisations have existed and therefore required some way of managing them, the separation in space and time between parts has been proportional. The separation in space is given by the geographical location of the parts around the world as it is known. This is nicely illustrated by a map of what the world looked like in the time of the Roman Empire. India and Rome, for instance, were far apart. It took a long time to travel between them and to transport anything between the two counties, including news and information.

This was the same for every society, for every administration, and for every organisation. There was thus for a long time no difference between the transportation of physical objects (including people) and information. But while the former could not easily be made faster, at least not until the invention of the steam engine, the latter could. An early solution to speed up the transmission of information was a system of beacons – fires lit on

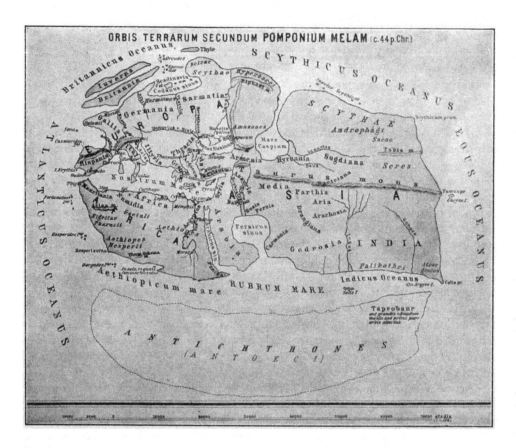

Figure 4.4 The Ancient world.

hills or high places for signalling over land that enemy troops were approaching in order to alert defences. Beacons have been used for thousands of years over much of the world, either as simply (binary) signals or as codes, as when several fires were lit in a specific pattern. A major step forward was the optical telegraph invented in 1792 by Claude Chappe in France. But it was not until the invention of the electrical telegraph, when human vision was replaced by electrical currents, that the speed of transmission really started to increase. The electrical telegraph overcame the dependence of the optical telegraph on visibility and meant that messages now could be sent over long distances – even crossing oceans – around the clock and that the time of transmission was negligible.

As long as there was proportionality between the transportation of physical objects and the transmission of information, it was only possible – but also only necessary – to manage and control the local surroundings. From a control engineering point of view the long distances would delay feedback (information) so much that control became impossible. When that no longer was the case it became possible – and indeed also necessary – to take into account and be aware of events that happened far away. The local focus that matched the "natural" limitations or capacities of the human mind therefore became insufficient. Humans have tried to handle the need to consider an increasing number of additional matters through various simplifying strategies, but the inevitable consequence of that is a fragmented view of the world. This is the serious challenge to control that cannot be solved by imagining that the world is simpler than it actually is.

Consequences of incomplete knowledge

The law of requisite variety is not only of relevance for the control or management of a system but also for whether a system can be analysed – for instance to assess risks and opportunities. That this must be so is obvious if we consider the opposite. If we do not have a clear description or specification of a system, or if we do not know what goes on "inside" it, then it is clearly impossible to effectively understand it and therefore also to investigate accidents or assess risks and opportunities. This lack of knowledge may apply either to how the system works (i.e., the "inner" mechanisms or comprehensibility) or to what the consequences of specific actions and interventions will be.

While we may entertain the hope that complete knowledge in principle is possible for predominantly technological systems (barring the vagaries of software), there is little reason for such optimism in the case of socio-technical systems. Here partial or complete ignorance is a fact of life because it is impossible fully to define or identify the parameters that are necessary to describe or model the system's performance, let alone to describe them in a realistic space-time continuum. The main reason for this is not that there are too many parameters, but rather that the systems are dynamic – i.e., that they continuously change – hence intractable.

- True ignorance means that it is impossible both in practice and in principle to get complete information about how the system functions or even about how it is structured. In terms of outcomes, this corresponds closely to the category of unexampled events (Westrum, 2006) – i.e., something that has never happened and for which, therefore, there can be no experience.
- Pragmatic ignorance means that it has been decided that for some things it is not necessary to know much – or even anything – about them. Such decisions are, however, always relative rather than absolute. They reflect the judgement that the

benefits of spending additional effort and time to find out something are of marginal value. This can be seen as representing a kind of efficiency-thoroughness trade-off, where thoroughness yields to efficiency (Hollnagel, 2009). If pragmatic ignorance is habitual, it soon becomes indistinguishable from complacency.

• Finally, there is wanton ignorance, which means that it is decided *a priori*, rather than by a trade-off, that something is devoid of interest. This corresponds to Merton's notion of an "imperious immediacy of interest," which denotes instances where a decision maker's paramount concern with the foreseen immediate consequences excludes the consideration of further or other consequences of the same act (Merton, 1936).

Losing and regaining control

If control is lost partially or completely, the "logical" conclusion, from the point of view of the law of requisite variety, is to increase the variety of the controller, as suggested by Conant and Ashby (1970). But there is also another solution – namely, to limit the variety of the system to be controlled. The net outcome will be that the variety of the controller matches or exceeds the variety of the system to be controlled. The solution of reducing the variety of the system is frequently used in the guise of standardisation and compliance. If work can be standardised and if people can be made to comply with guidelines and procedures, then the variety of the system can in principle be reduced until it matches the variety of the controller. In other words, eliminating or dampening the variability of the system makes it easier to control. This has historically been the preferred solution in warfare, in games, and in competitive situations generally. It is also used when authorities try to regulate companies, when companies try to regulate the workforce, and through multiple levels of the organisation where everyone as a rule tries to constrain the variability of those they have to manage or work with until they reach the proverbial sharp end, who can do nothing but comply. The downside is that it is impossible to guarantee that unexpected situations never will occur (cf. the earlier discussion of thermodynamic and anthropogenic entropy). When such situations arise, it may be the variety – or the ability to improvise and adjust – that in the end ensures that the situation is kept under control. Standardisation is therefore a solution that should be used with considerable care because the disadvantages more often than not outweigh the advantages.

It is an interesting consequence of the LoRV that attempts to reduce cognitive overload or complexity are futile. The reason is that this inevitably will lead to a simplified understanding and hence a simpler model for the controller. This will be insufficient unless the variety of the real world is actually also reduced so that the model and reality match. To be successful we must somehow be able to constrain the variety of the environment and the system to be controlled. But that in turn requires a thorough understanding. We are therefore caught in a nasty circle that has no solution. The only way out is to do the opposite – i.e., develop a model of the world that has sufficient variety or richness.

Change management myths

Since effective management requires that the system is understood, and preferably in a functional rather than structural sense, the question is how we can understand something. The classical approach to understanding something is decomposition, breaking the whole into parts. This can be found in Democritus's atomism as well as in Toyota's principle of the five whys. The principle of decomposition has unarguably been extremely successful both as the basis for our understanding of the natural world and as the basis for building ever more intricate systems, large and small. Decomposition has been particularly useful in explaining how something happens and is the dominant principle in most of the methods that are used across society. It is also how we try to understand hard problems, from consciousness to the meaning of life, the universe, and everything.

Socio-technical systems are unfortunately mostly intractable and cannot be understood by applying the principle of decomposition. One reason is that they do not function according to the assumptions of causality and linearity. A second reason is that they are dynamic rather than static and hence change all the time. The third and perhaps most important reason is that they are active rather than passive – i.e., they try to understand us as we try to understand them.

In the classical economics of Adam Smith, the world was either static or changed so slowly that it could be considered stable. This meant that changes and development could be contemplated by themselves, without the need to take into account the dependency on the environment. This has led to two important myths that are at the core of change management. But, being myths, they are both wrong.

The first myth can be called the *ceteris paribus* or "other things being equal" principle. The assumption of *ceteris paribus* – that other things are equal – is fundamental in Western thinking and essential for the established scientific method and the empirical testing of hypotheses. Carrying out a test of predictions from a hypothesis presumes that the planned intervention – the independent variable – is the only thing that affects how the system performs. Otherwise it would be impossible to conclude that the outcomes – the changes to the dependent variable – were the results of the independent variable. When we plan to do something, we must be reasonably certain that the assumptions on which the planning is based remain valid or at least reasonable during the time it takes to make a change. In particular we must be certain that nothing else happens that might affect the outcome. In scientific studies, whether of physical or behavioural systems, this can be achieved – or at least the attempts can be made – by following rigorous scientific methods and paradigms. But it is practically impossible to do this for complex socio-technical systems.

At the very least it is necessary to ensure that the understanding at the beginning remains valid during the change or study. Since conditions and the environment are constantly changing it is necessary from time to time to check whether the understanding is correct. This is sometimes referred to as *truth maintenance*. (Truth maintenance is a technique developed by artificial intelligence researchers in the 1980s to maintain consistency between the knowledge in a knowledge base and the current knowledge – i.e., the actual state of the world being reasoned about.) The longer it takes to make or implement a change, the less likely it is that the *ceteris paribus* principle is fulfilled. If it is impossible to guarantee that there will be no other changes during a planned intervention – i.e., to guarantee that *ceteris paribus* is upheld – then the best solution is to make the steps so small and the duration of the change so short that it is reasonable to assume there is no change in the conditions or assumptions. Unfortunately, few changes to how an organisation works are of that nature. Even worse, it would require that a change have only local effects in a system, and certainly no other effects than those that were foreseen. This is the subject of the second myth, which is called the substitution myth.

The substitution myth represents the assumption that artefacts can be value neutral in the sense that the introduction of an artefact into a system has only the intended (and no unintended) effects. The basis for this myth is the concept of interchangeability, as used in the production industry, where it was the basis for mass production even before Henry Ford. Thus, if we have a number of identical parts, we can replace one part with another without any adverse effects.

While in practice this holds for simple artefacts – the level of nuts and bolts – it does not hold for complex artefacts. (Indeed, it may not even hold for simple artefacts. To replace a worn-out part with a new one means that the new part must function in a system that itself is worn out. A new part in an aged system may induce strains that the aged system can no longer tolerate.) A complex artefact – using the term in a broad sense that also includes procedures and rules of communication – will require some kind of interaction, either with other artefacts or subsystems or with users, and hence can never be value neutral: "[n]ew tools alter the tasks for which they were designed, indeed alter the situations in which the tasks occur and even the conditions that cause people to want to engage in the tasks" (Carroll & Campbell, 1988, p. 4). In other words, introducing such an artefact into a system will cause changes that may go beyond what was intended and be unwanted (Hollnagel, 2003).

5 Fragmented change management

Introduction

The need to manage change exists for many reasons: because change is desired to improve performance, because change is needed to compensate for degradation due to ever-increasing entropy, because change is needed to recover from a disruptive event, because change is needed to counter intentional challenges or threats from an adversary, and because change is necessary to cope with the irregularities and unpredictability of the surroundings to ensure that an organisation can continue to function as it should. The essence of change management is finding ways to "move" individuals, teams, and organisations from a current state to a desired future state. Change management is usually seen as involving the problems of how to manage people in changing contexts: how to get the employees in a company involved in changes, how to get them to contribute to changes, and how to ensure that they accept the results. The main literature on change management, academic as well as practical, therefore concentrates on issues such as communication and motivation, on how to produce an actual change and how to make it "stick." Numerous websites and services offer advice on the principles necessary to lead change management but few provide practical details on change management seen as the control of a process.

The focus of synesis is on change management understood as a "voyage" of an organisation from the current to a new position or target and hence as controlling an organisation's movement. This must obviously encompass how to involve the members of an organisation in the change – the salaried employees as well as the management and board levels, forgetting neither the temporary employees nor the subcontractors. The focus in this book is nevertheless on how to manage an organisation as a complex system and specifically on how to avoid or reduce the consequences of the omnipresent fragmentation that characterises current practices and in many ways also provides the basis for them. Change management is usually pursued with a single issue in mind, be it productivity, quality, safety, or reliability, even in organisations where several, or all, of them are essential. Furthermore, each issue is typically managed by focusing on a few individual aspects or factors rather than by trying to deal with it in its entirety.

Managing productivity

The purpose of managing productivity is to ensure that something *does* happen and that there will be an increase in outcomes of a certain type. The same is the case for quality and reliability, but not for safety. Here the purpose is traditionally to ensure that something does *not* happen. Although there are important differences among the four, they are nevertheless similar in the sense that they focus on individual issues looked at in isolation.

Productivity is a measure of the efficiency of production. It is usually defined as the ratio of actual output (production) to what is required to produce it (inputs) and is often measured as a total output per unit of a total input. Productivity must be managed to maintain or improve the ratio since this most often is the principal source of income for an organisation. Productivity growth is important because more real income improves an organisation's ability to meet its obligations to customers, suppliers, workers, shareholders, the public, and authorities. Income also provides the resources required to maintain an acceptable level of safety and quality and to ensure reliability. Even non-profit organisations have a need of resources to safeguard their continued existence, although some of these may come from outside funding.

In many organisations the management of productivity relies on some form of control system. This unavoidably involves procedures for collecting, analysing, and reporting productivity data, often assigned to a unit such as the accounting department. Since production – unlike safety – always has been explicitly designed and planned, the process will be known and reasonably well described. It is it therefore possible to specify the relevant measurements and indicators and also to propose the interventions – the control actions – that are necessary and should be sufficient to bring about the desired changes. In terms of the voyage metaphor, it is known how an organisation functions and how the functioning can be made to go in a desired direction. Managing productivity is, however, more than just fine-tuning a complicated piece of real or metaphorical machinery and depends critically on managing people in changing contexts. This is done by identifying a set of critical or important obstacles (e.g., failures in communication, poorly defined task prioritisation, or a clash of expectations concerning deadlines and deliveries), by improving internal and external communication, by setting clear goals and deadlines, by effectively managing time and resources, by monitoring the execution of deliverables, and by preparing for unplanned events, hence partly overlapping with quality and safety management.

Faster, better, cheaper

A particular version of managing for productivity is the principle that projects should be completed faster, better, and cheaper. The faster, better, cheaper (FBC) strategy was adopted as a systems development philosophy by the NASA administration in the mid-1990s in line with the Clinton administration's approach of doing more for less (Paxton, 2007). The FBC philosophy was initially successful but was later seen as the reason for several spectacular failures. One of these was the loss in 1999 of the Mars Climate Orbiter and Polar Lander due to a discrepancy between the non-SI units used by the ground-based computer software, supplied by Lockheed Martin, and the SI units used by a second system, supplied by NASA in accordance with the software interface specification.

One problem with the FBC principle as a way of managing productivity is that of optimisation in a zero-sum world. The principle might possibly make sense if faster, better, and cheaper were recognised as three separate criteria since it would then be obvious that they are in competition. This problem has been treated by linear programming and solved within the constraints that exist there. But a solution always requires that something gives; i.e., it is impossible to optimise all three dimensions at the same time. The criteria of faster, better, and cheaper are not independent of each other, and lumping them together as FBC does not make them so. (Complex problems do not become simple by pretending that they have simple solutions.) The engineering adage "faster, better, cheaper: pick any two" acknowledges that while it may be possible to optimise $n-1$ out of n criteria at the same time, it is impossible to optimise all n. Lumping them together as FBC unfortunately does not unify them; it only makes them appear parallel rather than sequential. Their couplings and interrelations are still unspecified and implied rather than explicit.

Another problem with FBC is that the optimisation refers to the parts and not to the whole, or, rather, it is assumed that the optimisation of the whole, of the system as such, is a resultant outcome of optimising each part by itself. The optimisation of the whole is, however, more likely to be an emergent outcome, which means that one cannot achieve optimisation in the typical engineering way.

A further complication that is usually overlooked is that changes or improvements do not happen equally quickly in all parts of the system; they are usually asynchronous rather than synchronous. If we consider just the four issues used here – productivity, quality, safety, and reliability – then it is clearly unrealistic to assume that they develop on the same time scale or have identical dynamics. Trying to make them all faster, better, or cheaper – or even worse, faster, better, *and* cheaper – at the same time is clearly doomed to failure. Since the four issues, furthermore, are mutually dependent or tightly coupled rather than independent, it is hardly a surprise that a simple-minded approach such as FBC will fail sooner or later.

Managing quality

The importance of ensuring the quality of manufactured products has always been recognised, not least since mass production became established after the Industrial Revolution in the late 18th century. It received a significant boost with Shewhart's book in 1931, as described in Chapter 2. Shewhart showed how quality could be managed by means of statistical process control, which leads to the development of specialised areas of expertise, such as quality planning, quality control, and quality improvement. It nevertheless took sixty years before the term *quality management system* (QMS) was invented (according to the Wikipedia entry on QMS, the term was suggested in 1991 by Ken Croucher, a British management consultant working within the IT industry). It is today enshrined in the ISO 9001:2015 standard, where it is described as necessary for an organisation that "needs to demonstrate its ability to consistently provide products and services that meet customer and applicable statutory and regulatory requirements, and that aims to enhance customer satisfaction through the effective application of the system, including processes for improvement of the system and the assurance of conformity to customer and applicable statutory and regulatory requirements" (ISO, 2015). The standard further points out that all the requirements "are generic and are intended to be applicable to any organization, regardless of its type or size, or the products and services it provides."

The purpose of managing quality is to ensure acceptable outcomes from a production line or a service function, with due consideration of the constraints imposed by efficiency, effectiveness (cost), sustainability, etc. The quality of products or services is a key component to a company's success, and QMS serves to align diverse aspects of a company with the single purpose of delivering acceptable products and services to customers. But quality in this way becomes managed by itself by a separate organisational unit, with little if any consideration of the other issues. Quality management has also developed its own set of methods or techniques, such as total quality management (TQM), Six Sigma, and

Kaizen, which are used in varying ways according to the requirements of an organisation. There is regrettably no single technique for quality management that comprises both the production of the product or service and an organisation as a socio-technical system. It has, predictably, been proposed that there is something called a quality culture, defined as "[a] culture of quality . . . in which everybody in the organization, not just the quality controllers, is responsible for quality" (Harvey & Green, 1993). This definition is not too different from the popular definition of a safety culture as the collection of beliefs, perceptions, and values employees share in relation to risks within an organisation. And it is probably just as effective.

Similar to production management systems, the management of quality also involves a number of distinct aspects, which include defining quality objectives, producing a quality manual, controlling documents, defining organisational roles and responsibilities, sampling and managing data, and ensuring continuous improvement. These are, in practice, addressed individually rather than together as the remit of a quality department even though many of them are similar to aspects defined for productivity or safety.

Products and services

There is a significant difference between managing the quality of products and managing the quality of services. For products it is essential that each product is like the other. This applies to material items (consumer goods, gadgets, cars) as well as consumable items such as vitamin pills and ice cream. If I buy something, whether to use it or consume it, I expect the same standard each time and accept that others who buy will also get the same standard.

In today's society and world the production of goods is rivalled by the production of services. According to the U.S. Bureau of Economic Analysis, the services sector in 2009 accounted for 79.6 percent of U.S. private-sector gross domestic product (GDP). Services must also be of high quality, but that does not mean that they can be standardised in the same way as products. When I want a high-quality treatment, I want individualised treatment, one that I am satisfied with, but not the same treatment that everybody else gets. Even when I buy an expensive standardised product, such as a car or a smartphone, I want individualised service. This is even more obvious in the case of more complicated services such as treating patients in a hospital. From the patient's point of view, the treatment must be individualised (also because people are different and rarely suffer from identical illnesses). And if I hire someone to organise a social event, I don't want a standard product; I want an individual product. But this means that quality cannot be assured in the same way as for manufactured products. It is a question of the quality of how an organisation and services function as much as a question of the quality of a produced item or artefact.

Managing safety

While the purpose of managing productivity and quality is to ensure that something *does* happen, the purpose of managing safety is to ensure that something does *not* happen or that there are no outcomes of a certain type. The purpose of safety management is, in practice, to reduce the number of unwanted outcomes from an activity or operation and hence to manage the absence rather than the presence of safety (cf. the discussion of Safety-I and Safety-II in Chapter 2). A safety management system (SMS) can be described as "a suite of systematic, explicit and comprehensive processes for managing safety risks.

An SMS provides management with a directed and focused approach to safety with a clear process for setting goals, planning, and measuring performance" (International Transport Forum, 2018). The term *safety risk* may seem a little puzzling since safety semantically and pragmatically ought to be the opposite of risk. It has nevertheless become part of the standard safety management vocabulary and has been defined in analogy with risk as "the quantification – expressed in terms of predicted probability and severity – of the potential consequence(s) of a hazard taking as reference the worst foreseeable (but credible) situation" (Maurino, 2017).

The emphasis of safety management is to respond to hazards, usually after they have manifested themselves through accidents, as well as to identified risks, including, as far as possible, the latent conditions that may become active risks under unfavourable circumstances. In both cases the purpose is to eliminate the hazards or risks, or at least to reduce them to an acceptable, which in practice means affordable, level. Safety management is therefore focused on individual hazards or risks and addresses these in a standardised manner, such as accident analyses or risk assessments. The predominant approach can be characterised as "find-and-fix," which means that the focus is on identifying the presumed sources or causes of what could or did go wrong. Because each problem is addressed and solved on its own, there is little if any continuity of either management or learning. This is not least so for industries that have relatively few accidents and that consequently consider themselves very safe or even ultrasafe. Indeed, the lofty target of zero accidents will, if it is ever achieved, effectively remove any basis for further learning.

Managing reliability

The fourth issue described in this book, and also historically the last (cf. Chapter 2), is reliability. Reliability means that something – or someone – can be trusted to perform consistently well. A reliable system is often defined formally as one that performs according to specifications. Since reliability refers to the performance of something or someone, there is traditionally a distinction between the reliability of technology, humans, and organisations. Each is addressed by its own subdiscipline, known as reliability engineering, human reliability assessment, and high reliability organisations, respectively. The three subdisciplines regrettably have little in common with each other.

It is clearly important that we can rely on the technological parts of a system to function as specified when needed. As described in Chapter 2, reliability engineering started in the early 1950s with the need to ensure the reliability of military equipment but has since been adopted wherever the reliability of technical components and equipment is important – which today means practically everywhere. Reliability is often expressed as the probability of failure-free operation for a specific period of time in a specified environment (i.e., under nominal conditions, since nothing can be guaranteed to function under all conditions). There is therefore an obvious relation between quality and reliability. If a piece of equipment is lacking in quality or is of low quality, then it is unlikely to be reliable. But quality focuses on manufacturing defects during production (possibly extended to the warranty phase) and is therefore related to the control of lower-level product specifications. Reliability must cover the whole life of a product, engineering equipment, or system from commissioning or birth to decommissioning or death; quality hence needs to consider the assessment of reliability and failure rates over years, decades, or even longer. As experiences from managing productivity and quality were already available, the management of reliability came to include many of the same components or foci. The recommendations for a reliability management system emphasise such items as strong leadership, effective communication, data collection and analysis systems, procedure, document and knowledge management support systems, and even reliability culture management.

Reliability is a complex rather than simple phenomenon. Productivity can be measured or assessed by counting the output for a given period or calculating the ratio between input and output. Quality can likewise be assessed directly either by sampling or by direct inspection of items. Both are therefore relatively simple. Safety, of course, is a little more problematic because the focus is on the absence rather than the presence of safety. Since it is impossible to count or measure how often something does *not* happen, the solution has been to count adverse outcomes even though they only mark the absence of safety. In all three cases it is also possible to intervene or make changes more or less directly. Reliability is, however, quite different. It is first of all not something that can be recognised immediately since it involves performance over an extended period of time. (Productivity, quality, and safety issues can be recognised very quickly.) It is also based on a comparison with identical pieces of equipment or similar processes, and a single instance is not sufficient to claim that reliability is low. For the management of reliability it means that there are few cases of direct connections between interventions and outcomes.

Human reliability assessment (HRA)

For a long time it seemed that reliability engineering had solved the problems or at least been sufficient to bring about an acceptable – or affordable? – level of reliability in technical systems. Since reliability was defined operationally as the probability of failure-free operation for a specific period of time in a specified environment, the calculation of the failure probability became the common measure for reliability, and reliability was expressed as the failure rate, or, rather, as $1 - p$(failure). (NASA, for instance, defines risk as the probability of failure, called Pf, and reliability as the probability of success, called Ps, where $Ps = 1 - Pf$.) This approach served the community well until the partial meltdown of reactor number two of the Three Mile Island Nuclear Generating Station near Harrisburgh, Pennsylvania, happened on March 28, 1979. The accident made it glaringly obvious that it was not enough to address the reliability of technological equipment – the human "component" also had to be taken into account. The response was the subdiscipline known as human reliability assessment (HRA). Human reliability was initially defined as "the probability that a person (1) correctly performs some system-required activity in a required time period (if time is a limiting factor) and (2) performs no extraneous activity that can degrade the system" (Swain & Guttmann, 1983, pp. 2–3).

The concept of human reliability created considerable controversy from the very beginning and can still be the subject of heated debates. There is no reason to delve into this here since the volume of books and papers is quite considerable. The more interesting issue is, of course, how to manage the reliability of humans. Unlike technical systems humans are not designed, and despite 140 years of psychology as a scientific discipline we know only in rather general terms how humans think and even less about how we can manage that to ensure reliable performance. This clearly presents a problem for the management of reliability in general since all systems are socio-technical systems in one way or another and hence include and depend on the reliability of human performance on many levels.

High reliability organisations (HRO)

When the issue of human reliability had been solved – or at least addressed – it was assumed that the problem of reliability management had been brought under control, the practical problems of managing human reliability notwithstanding. But it soon became clear that there was an additional factor at play – namely, the organisation. Just as technology and humans could perform unreliably, so, too, could organisations. The problem had been identified by Charles Perrow's seminal book *Normal Accidents: Living with High-Risk Technologies*, published in 1984, which argued that many failures relate to organisations rather than technology or humans. The problem with how well organisations functioned was clearly demonstrated in 1986 with the Space Shuttle Challenger disaster and the disaster at the Chernobyl nuclear power plant in Ukraine, at that time part of the Soviet Union. This was just two years after the publication of *Normal Accidents* and only seven years after human reliability had become an issue. While investigations into the two disasters, among several other accidents, focused on what had gone wrong and, incidentally, led to safety culture being proposed as a panacea, the interest of a group of researchers at the University of California, Berkeley, focused on organisations that were able to function "error free" in conditions where "normal accidents" would have been expected to happen.

The intensive research into high reliability organisations (HRO) has produced a good understanding of the five qualities that characterise such organisations, as described in Chapter 2. What unfortunately is lacking are more practical guidelines or principles for how to manage organisations to ensure that they can perform reliably. Organisational reliability, like safety, has tended to look at catastrophes and failures and provide advice on organisational strategies that can be used to respond to "dysfunctional characteristics and processes" (e.g., Roberts, 1990) – in other words, how to manage the lack of organisational reliability rather than how to manage reliability.

Fragmentation in change management thinking

The characterisations of how change management has developed with regard to the four issues show that the thinking is fragmented in several ways. Productivity started with the organisation of work and is still very much focused on that. In the beginning, scientific management was focused on the individual worker, who was given economic incentives to work as prescribed, as a machine. But today the emphasis is mostly on the organisation, and that in an even wider sense since the dependencies have grown with the vertical and horizontal expansions. Quality started with the product, acknowledging that it depended both on the human worker and the organisation of work. It has very much stayed there, although it is slowly spreading to the organisation in the common enthusiasm for culture. Safety also started with a focus on the human but, reluctantly, after 1986 moved on to the organisation and safety culture. Safety management has, oddly enough, never dealt with the inanimate as such. When this happened the issue was reliability rather than safety, although there was a clear relation via risk. Safety, as Safety-I, may, however, eventually have to include the inanimate as automation increases and AI is introduced in the optimistic hope that it will solve all the hard problems. Finally, reliability started with technology, then spread to humans and then to the organisation but as three separate strands rather than as a unified concept.

The main reason for the fragmentation in change management thinking is unquestionably the different beginnings of the four issues as well as their differences in emphasis.

Each issue was from the start managed in its own way and soon developed its own traditions. It is hardly surprising in the case of the first issue, productivity. Neither is it surprising that this laid the basis for thinking about the organisation of work. But it is somewhat surprising, at least in hindsight, that the issues that came later – particularly quality and safety – did not consider or acknowledge the relation to productivity. Instead, they addressed their specific issue or concern in isolation, creating new specialised departments, new models, new methods, and new cultures. This is even more surprising because many of the main ideas were transferred from one issue to the next in a rather "mechanical" way, as described in the following:

- Task decomposition was introduced for the productivity issue but has been used by the other three as well. Decomposition is generally used to make it easier both to understand something and to manage something successfully. Productivity also embraced the idea of Work-as-Imagined, although the term did not exist at the time.
- The elimination of waste also comes from productivity and later became the rationale for Lean.
- Specialisation and standardisation were also introduced for productivity, primarily to improve effectiveness. For quality, standardisation was a means of reducing overall variability. For safety and reliability, specialisation and standardisation can be found in the emphasis on procedures and compliance.
- The use of statistics, as in statistical process control, came from quality. Statistics, in the sense of frequency and probability, is also an important part of reliability but is less important with regard to safety.
- Linear cause-effect analysis comes from safety, most obviously in the case of root cause analysis. Cause-effect analysis is also part of quality, but only in relation to assignable causes. Indeed, the concept of chance causes actually means that the "normal" variability of processes is seen as not being of interest.

These legacies have together created a fragmented view that has become second nature to change management and is therefore no longer noticed. There are additional forms of fragmentation, but to understand them it is necessary to go through some of the main approaches to change management.

The necessity of change

Returning to the voyage metaphor, change management requires the competence to consistently set a course, adjust it whenever needed, and ensure that it is followed until the set waypoint or target has been reached. It is first of all important that change management is systematic and that it follows a suitable plan of action. This means that there must be sufficient time to plan how the changes should be brought about in detail. Change management should not be simple reactions but rather must be the outcome of an ongoing process that requires – and deserves – time and expertise. Neither should change management be an isolated activity that is separate from the normal functioning of an organisation. Indeed, the rationale for change management is to support, sustain, and improve the normal functioning. It is therefore hardly surprising that there are many suggestions for how change management can best be accomplished, not least in the modern management literature. The following sections will look at three major suggestions in chronological order.

From specification-production-inspection to plan-do-study-act

The prototypical approach to change management in quality control is the so-called Shewhart cycle described by Walter A. Shewhart (1939). The Shewhart cycle consists of three steps – specification, production, and inspection – that are the basis of all production but are specifically important in quality control. Shewhart argued that they should be seen as a cycle rather than as a simple sequence:

> These three steps must go in a circle instead of in a straight line. . . . It may be helpful to think of the three steps in the mass production process as steps in the scientific method. In this sense, specification, production, and inspection correspond respectively to making a hypothesis, carrying out an experiment, and testing the hypothesis. The three steps constitute a dynamic scientific process of acquiring knowledge.
>
> (Shewhart, 1939, p. 45)

Shewhart here refers to the scientific method described by the English philosopher and statesman Francis Bacon in the *Novum Organum*, published in 1620. Francis Bacon was trying to give a description of how humans can come to knowledge and argued for a practical and empirical approach in contrast to the rational and theoretical approach advocated by the contemporary French philosopher and mathematician Rene Descartes. The scientific method begins by formulating a question or stating a problem; this leads to the formation of a hypothesis, followed by an experiment to test the hypothesis; the results are then measured and recorded; and finally the data are analysed to determine whether the hypothesis was correct. In a shorter version the steps are formulating a hypothesis, carrying out an experiment, and evaluating the outcome of the experiment. The similarity between the two approaches is clear to see from Figure 5.1. And it becomes even more obvious if the three steps are called to plan, to do, and to study.

The Shewhart cycle originally had only three steps but soon got a fourth – namely, to act. This is a logical consequence of the reason for inspecting and evaluating the outcomes in the first place. An inspection or evaluation can either conclude that the product

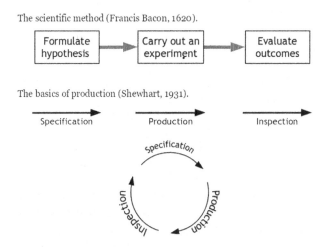

Figure 5.1 The scientific method and change management.

or outcome is of acceptable quality and that it corresponds to the agreed standards, in which case actions should be taken to ensure that this continues to be the case, or it can conclude that the product is not of acceptable quality, in which case it will be necessary to do something, to adjust or modify the production process in some way. Shewhart's book from 1931 was entitled *Economic Control of Quality of Manufactured Product*, which clearly implies that the conclusions of the evaluation should be the basis for appropriate corrective actions. To manufacture or produce something, a physical artefact or a service, is by its very nature a continuous activity or must at least include sub-stages of continuous activities. The fourth step thus establishes a loop or cycle, a "circle instead of in a straight line," as Shewhart pointed out (op. cit.). By contrast, the process of acquiring knowledge, as described by the scientific method, is not continuous in the same sense even though it is accumulative in the long run.

Today the four-part cycle is commonly known as the PDSA cycle, where the letters represent plan, do, study, and act, respectively. A good account of how this happened, and also why the intermediate version called plan-do-check-act (PDCA) fell from grace, has been provided by Moen and Norman (2009). PDSA is also often called the Deming cycle, after W. Edwards Deming, who is recognised as the father of modern quality control.

The origin of the PDSA cycle was quality control of manufactured products, where the same process is carried out repeatedly, which makes it possible to make small changes again and again. In that context the iterative use of the PDSA cycle was seen as instrumental in solving identified problems in a step-by-step fashion. Moen and Norman (2009) pointed out that iteration is a fundamental principle of both the scientific method and PDSA and that repeating the cycles brings users closer to the goal. But today the PDSA cycle is also widely used as the recommended, or even mandatory, approach to change management in contexts where the processes are socio-technical rather than technical and therefore less stable and regular – for instance in health care settings (e.g., Donnelly & Kirk, 2015). The ISO 9001:2015 standard thus notes that "the PDCA cycle can be applied to all processes and to the quality management system as a whole." In such cases there is often a need to see more dramatic improvements to justify the efforts and costs necessary to make a change. Although the PDSA cycle was developed for a completely different context, it is nevertheless the approach that is usually recommended – for instance by the Institute for Healthcare Improvement (IHI). (IHI is an independent not-for-profit organisation that helps improve health care throughout the world.) IHI's website provides the following recommendation for change management (see also Figure 5.2):

> Once a team has set an aim, established its membership, and developed measures to determine whether a change leads to an improvement, the next step is to test a change in the real work setting. The Plan-Do-Study-Act (PDSA) cycle is shorthand for testing a change – by planning it, trying it, observing the results, and acting on what is learned. This is the scientific method, used for action-oriented learning.
> (Institute for Healthcare Improvement, 2019)

The IHI approach leaves unsaid what the initial motivation is – i.e., what the reasons are for initiating a change. In the Shewhart cycle and the original PDSA the concern was to ensure stable product quality. This is, of course, to some extent also the case for health care systems, but in addition to that the aim may also be to standardise treatments (which does not always have improved quality as a result), to eliminate waste, or to make more

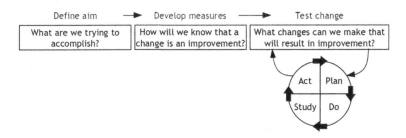

Figure 5.2 IHI Model for Improvement.

radical changes. In these cases it cannot be taken for granted that PDSA is the best way of accomplishing the goal.

In the context of this book, PDSA is suited to the management of stepwise improvements rather than to change management. It carries with it the assumption that the improvement is for an inert or passive system that does not change by itself. That was the case for manufacturing in the 1930s, but it is not the case for the socio-technical systems and complex organisations that must be managed today.

Action research and the spiral of steps

The Shewhart cycle, PDSA, the Toyota Production System, and several other variations on the same theme were intended to be used to ensure an acceptable level of quality of manufactured products. Although a factory, an assembly line, and a production facility are all clearly socio-technical systems, the approach these methods represent tacitly assumes that the changes are directed at the technical parts of the system rather than the social (human) parts. Social systems are, however, by their very nature different from technical systems, and changes to social systems must acknowledge and respect that. It is still the case that the basic principles of the scientific method hold – formulating a hypothesis, followed by experimenting, and concluded by evaluating. But it is essential to recognise that social systems are different from technical systems in nearly every possible way. They are underspecified and forever changing (since they actively try to compensate for increasing entropy), and they try to anticipate what may happen and what others may do (using newer terminology, one might say that they are able to perform in a resilient manner). It can therefore not be taken for granted that the Shewhart cycle or its derivations can be used to manage changes in social systems. Fortunately, there are alternatives.

The main inspiration for change management in social systems is to be found in action research. This is the common name for a research methodology that was developed in the 1940s to bring about transformative change through the simultaneous process of taking action and doing research and linking the two together by critical reflection. The agenda for action research was formulated by Kurt Lewin, a German-American psychologist who is generally acknowledged as the founder of social psychology. Action research is an iterative process in which research leads to action and action leads to evaluation and further research. Lewin introduced it in the following manner:

> In the last year and a half I have had occasion to have contact with a great variety of organizations, institutions, and individuals who came for help in the field of group

relations. . . . These eager people feel themselves to be in a fog. They feel in a fog on three counts: 1. What is the present situation? 2. What are the dangers? 3. And most importantly of all, what shall we do?

(Lewin, 1946, p. 201)

It is not difficult to see the similarity between the three questions Lewin identifies here and the three steps of the scientific method and the Shewhart cycle – and also the similarity to the voyage metaphor. It becomes even more obvious when Lewin explains how action research "proceeds in a spiral of steps each of which is composed of a circle of planning, action, and fact-finding about the results of the action" Lewin (1946, p. 206). Each of the three main parts in the spiral of steps can be described as follows (cf. also Figure 5.3):

The first step then is to examine the idea carefully in the light of the means available. Frequently more fact-finding about the situation is required. If this first period of planning is successful, two items emerge: namely, "an overall plan" of how to reach the objective and secondly, a decision in regard to the first step of action. Usually this planning has also somewhat modified the original idea.

(Ibid., p. 205)

The first step thus corresponds to the "plan" of PDSA. The second step is essentially the execution of the first step of the overall plan and therefore corresponds to the "do" of PDSA. It is followed by the third step, originally called reconnaissance or fact-finding, which has four functions:

First it should evaluate the action. It shows whether what has been achieved is above or below expectation. Secondly, it gives the planners a chance to learn, that is, to gather new general insights. . . . Thirdly, this fact-finding should serve as a basis for correctly planning the next step. Finally, it serves as a basis for modifying the "over-all plan."

(Ibid., pp. 205–206)

Lewin's model, or approach, does not include the fourth step of PDSA, to "act." As the aforementioned quotation shows, the "act" is already included in the fact-finding step. In

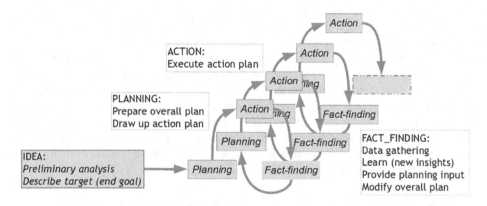

Figure 5.3 Lewin's Spiral of Steps model.

the IHI model, the meaning of "act" is to make a plan for the next step based on what has been learned from the test. What the PDSA cycle lacks, but what is present in the Lewin cycle, is the possibility of modifying or revising the overall plan. This makes the graphical rendering of the model less neat than the PDSA cycle but, by the same token, perhaps also more realistic.

It obviously requires careful planning to make a change to a technical system, such as a production process, but the system as such is inert or passive and therefore does not require any special preparations. It responds to the intervention, but it neither anticipates nor tries to resist it. However, it is a completely different story for changes made to social systems. People may resist the change, be indifferent to it, just go along while it happens and then later revert to what they did before, or enthusiastically embrace it. For those reasons the approach to change management in social systems must be different from the approach in technical systems. Lewin fully realised that and wrote:

> A change towards a higher level of group performance is frequently short-lived, after a "shot in the arm," group life soon returns to the previous level. This indicates that it does not suffice to define the objective of planned change in group performance as the reaching of a different level. Permanency of the new level, or permanency for a desired period, should be included in the objective. A successful change includes therefore three aspects: unfreezing (if necessary) the present level L^1, moving to the new level L^2, and freezing group life on the new level.
>
> (Lewin, 1951, p. 228)

A high-level approach to change in social systems thus involves the three steps Lewin called unfreeze, change, and freeze.

• The unfreeze step involves the preparation for the change. When a certain way of working has been in place for a while, habits and routine have naturally settled in. People throughout an organisation will have learned and agreed to do things in a particular way and may even have developed their own local culture. They have accepted that as efficient and therefore use it routinely without thinking much about it and rarely, if ever, consider whether there might be other, more efficient methods. The unfreezing is needed to get people to re-examine their day-to-day activities and habits as a precondition for setting the wheels of change in motion.
• The second step, change, has already been described by the spiral of steps method. Bringing about the change will probably take some time and involve a transition period. Relinquishing old habits is never easy and often not very welcome because it disrupts well-established routines. Progress may therefore at first seem slow for those who manage the change as well as for those who are subjected to it. A change process requires an investment by all in terms of time and allocation of resources, and the pay-off may be neither immediate nor easy to recognise. Most people, not least those who are supposed to manage, may also have very unrealistic ideas about how quickly an organisational change can take place and when the intended results can be seen.
• Once the change has been made, the final step is a freeze, or consolidation. Put differently, the change has intentionally disturbed an equilibrium. After the change has been made it may take some time before work settles down to a steady state. Yet it is necessary to ensure that the new way of working becomes the accepted

standard. It is only when a new equilibrium has been established that people throughout an organisation can reap the benefits of the change. This may take much longer than usually expected.

Observe–orient–decide–act (the OODA loop)

Change management can be seen as trying to control a transition from one state (type or mode of performance) to another. But change management can also be seen more directly as a kind of command and control. Command and control – usually referred to as C2 – is characteristically used in relation to military operations, where the terms are defined as "the creative expression of human will necessary to accomplish the mission" and "those structures and processes devised by command to enable it and to manage risk," respectively (Pigeau & McCann, 2002). The actions that are part of command and control are shown in Table 5.1. The concept has also, for a long time, arguably since the introduction of scientific management (cf. Chapter 2), been used in a business context but is today regarded by many as too traditional and inflexible.

Command and control, in the broader meaning of the terms, are necessary for making changes regardless of the context and therefore also for change management as it is described here. The "command" is the outcome or conclusion from studying and understanding the situation, interventions, or detailed plans for how to bring about a change. The "control" is the close tracking of how the interventions or actions develop to ensure that the aim is accomplished. This is nicely captured by the observe–orient–decide–act, or OODA, loop developed by USAF Colonel John Boyd (cf. Figure 5.4). (According to Brehmer [2005], the OODA loop was originally developed to describe fighter combat in terms of four activities, or stages, in order to understand why American fighter pilots were more successful than their adversaries in the Korean war.) The basic idea is to describe decision-making as taking place in a recurring cycle of observe–orient–decide–act.

> Yet, as we have seen, on one hand, we use observations to shape or formulate a concept; while on the other hand, we use a concept to shape the nature of future inquiries or observations of reality. Back and forth, over and over again, we use observations to sharpen a concept and a concept to sharpen observations. Under these circumstances, a concept must be incomplete since we depend upon an ever-changing array of observations to shape or formulate it.
>
> (Boyd, 1987, p. 4)

Table 5.1 Actions that are part of command and control

To command	To control
To create new structures and processes (when necessary)	To monitor structures and processes (once initiated)
To initiate and terminate control (this includes establishing the conditions for initiation and termination)	To carry out pre-established procedures
To modify control structures and processes when the situation demands it	To adjust procedures according to pre-established plans

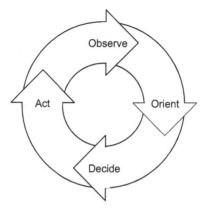

Figure 5.4 The OODA loop.

The essence of the four steps is that the combination of observation and orientation leads to a decision, which in turn leads to action. The purpose of "observation" is to understand what goes on by collecting as much information as practically possible. This is followed by "orientation," where the information from the first stage is organised and interpreted to yield a comprehensive understanding of the situation. Based on this, it is then possible to make a "decision," which determines what will or should be done. The fourth and final stage is the "action," where the chosen action or intervention is actually carried out. The consequences of the actions are observed, which means that the cycle is repeated. (In the specific context in which the OODA loop was developed, the observation could be what the opponent was trying to do in response to the action – for instance in aerial combat.)

In the context in which the OODA loop was developed, it was important to be as quick as possible, and definitely quicker than your opponent or enemy, who preferably should not have the chance to complete the loop as well and hence be kept running behind the facts. In this way the OODA loop differs from PDSA and Lewin's model since it assumes there is an active opponent, or even an enemy, and that events develop rapidly. That is perhaps why the business world has embraced the OODA loop rather than PDSA, as the latter basically deals with a non-adversarial environment that changes slowly.

OODA is adversarial change management, or just adversarial management, aiming to overcome external changes in a very dynamic environment where long-term planning is not really feasible. It is the trade-off between strategy, tactics, and operations. Change management is difficult when things in the surroundings happen faster than expected and faster than responses can be made. This rarely happens in relation to productivity and quality issues but may easily happen in relation to safety issues. Fast changes in the surroundings may also easily lead to a condition of IIO (cf. Chapter 3). Change management may paradoxically be even more difficult if the changes happen too slowly. If there is nothing to notice after the completion of the "do," is that because nothing happened or because it has just not happened yet? When something happens very slowly we have problems connecting the dots, and it can be difficult to recognise a pattern. Think of a melody that is played so slowly you hear only the individual tones, not the tune – or think of global warning.

Destroy and create

The description of what "command" implies (Table 5.1) includes both the creation of the objective or purpose of the activity ("to create new structures and processes") and their revision whenever needed ("to modify control structures and processes"). This is obviously essential in the context of military operations, where the OODA loop began, but is also relevant in adversarial business environments. In these situations it may be uncertain how the situation will develop and what the opponent may do. The spiral of steps includes something similar, in the form of an original idea or overall plan that is modified when needed, although in these cases there is no explicit opponent. By contrast, neither the Shewhart cycle nor the PDSA loop refers to the possibility of changing the initial target or the overall objective. This is not too surprising since both address systems that are assumed to be passive and hence harbour no intentions or initiatives of their own.

Another important difference from PDSA is that the concept – the result of "observation" plus "orientation" – is assumed to be incomplete. It must therefore be shaped and formulated through the OODA loop. This is similar to the ideas found in both Bacon and Lewin, but quite different from the PDSA cycle. The crucial difference between PDSA and the OODA loop is captured by the word *incomplete*. Instead of starting with a "plan," the OODA loop starts with an "observation," the main purpose of which is to "sharpen" the initial assumptions. The value of the change that is the final goal of the "act" is, in this way, gradually established through the loop rather than given from the start.

The need to revise the initial target or the overall plan is nevertheless essential for any kind of change management. Boyd was quite explicit about this when he wrote:

> To comprehend and cope with our environment we develop mental patterns or concepts of meaning. The purpose of this paper is to sketch out how we destroy and create these patterns to permit us to both shape and be shaped by a changing environment. In this sense, the discussion also literally shows why we cannot avoid this kind of activity if we intend to survive on our own terms.
>
> (Boyd, 1987, p. 1)

Boyd emphasised the need to "destroy and create" the understanding on which command and control, which in this context is equivalent to change management, is based. It is not enough to modify the overall plan when it needs it. It is also necessary to be willing to destroy it, which means to abandon it completely, and develop a new plan in its stead. He went on to note that

> the uncertainty and disorder generated by an inward-oriented system talking to itself can be offset by going outside and creating a new system. . . . [U]ncertainty and related disorder can be diminished by . . . creating a higher and broader more general concept to represent reality. . . . In this unfolding drama, the alternating cycle of entropy increase toward more and more disorder and the entropy decrease toward more and more order appears to be one part of a control mechanism that literally seems to drive and regulate this alternating cycle of destruction and creation toward higher and broader levels of elaboration.
>
> (Ibid., p. 6)

The "higher and broader levels of elaboration" correspond to an improved understanding of what is relevant for managing the change. Since an organisation that is being changed is an open rather than closed system, it is essential to understand where the boundaries lie and what determines the variability or uncertainty of the surroundings. A consequence of that may be a revision and expansion of the boundaries – in other words, a need to include more and more in the understanding of the "world." More will be said about this in Chapter 7.

In the world of today, at the start of the third decade of the 21st century, there is little stability in the organisations we need to change and manage, partly because they are active, partly because we unintentionally have set changes in motion (trade wars being a recent example) where we seem to understand neither their complexity or interdependency nor the rate at which they happen. Neither is there much stability in their surroundings. This lack of stability must be recognised as a fundamental premise for change management. Failing to do so will leave us with approaches that are more likely to fail than to succeed.

Fragmentation issues

The preceding sections have tried to characterise three of the most important change management approaches, none of them from the present century. A search of the web will quickly identify several others, but a closer look will soon reveal that they represent variations on already known themes rather than new themes.

Each of the three, and therefore also their many variations, is burdened by several types of fragmentation. This is rarely noticed in practice, not least because the fundamental fragmentation that is a consequence of focusing on a single issue – on productivity, quality, safety, or reliability but not on two or more of them together – emphasises the strengths of doing so and hides the weaknesses. There is little doubt that each approach to change management is, in practice, effective as long as the underlying assumptions are met. But the problem is that the apparent efficiency of the practice discourages any attempt to scrutinise the assumptions. As all methods they imply or carry with them an inevitable efficiency-thoroughness trade-off. Simply using them as they are and as we have been taught is efficient, and since everybody seems to do so the consequent lack of thoroughness is all too easily overlooked and forgotten.

Fragmentation of foci

One kind of fragmentation is due to the habit of focusing on a single issue or concern at a time. This is to a large extent due to the historical sequence of the four issues described in Chapter 2, the result of which has been four – or more – separate concerns, each with its own methods, theories, and models. Each focus has also created its own organisational silo, with specialised roles, procedures, and cultures. The fragmentation of foci can also be found in cases where multiple priorities have been introduced at the same time, such as the faster, better, and cheaper (FBC) idea. The fragmentation of foci is consistent with how the human mind works, as described in Chapter 3. The psychological reasons for fragmentation make it easier to focus on one thing at a time and leave the rest lurking in the background for later. When it ever so often becomes necessary to look at several things at the same time it can be overwhelmingly difficult to understand the dependencies and couplings both because the available vocabulary is poor and because there are only a few methods available. This was not a serious obstacle when the three main approaches were developed since the world then (1930s–1950s), using Perrow's (1984) terminology, consisted of loosely coupled systems with simple, linear interactions. The systems and organisations of today are, however, tightly coupled, with non-linear interactions bordering on entanglement. There are common-mode connections, interconnected subsystems, many feedback loops, and indirect information, all of which result in limited understanding.

Fragmentation of scope

Fragmentation is also a consequence of the way in which we attempt to understand larger systems by means of decomposition and to understand problems by breaking them into parts (sub-problems) that can be solved separately and by our continuing to do so until a level of elementary sub-problems is reached. The assumptions that the elements or constituents can be described or analysed individually is, of course, very convenient for change management since it means that changes can be implemented step by step. To the extent that the steps do influence each other, this is assumed to happen in a linear manner, in a known way, with a known speed. It is also assumed that the *ceteris paribus* principle is upheld. In other words, the whole is just the sum of the parts, or the whole can be expressed and understood as a (linear) combination of the parts. This is the case for understanding how technology and machines work (cf. the characteristic "exploded" drawings that show the relationship or order of assembly of various parts).

The fragmentation into parts assumes that systems can be described by their architecture, i.e., in terms of their structure. This requires that it is possible to clearly specify or recognise a boundary around the system, something that demarcates the system from its surroundings. It also requires that the same can be done on a smaller scale so that a system can be described as composed of subsystems. This is indeed the classical definition of what a system is:

> A system is a set of objects together with relationships between the objects and between their attributes.
>
> (Hall & Fagen, 1968, p. 81)

The boundary problem is, however, far from trivial, which is something that will be discussed in detail in Chapter 7. Suffice it to say that it may be both better and easier to describe systems in terms of their functions rather than in terms of their structures. This is partly due to the problem of intractability but is also a recognition of the fact that systems – organisations – are dynamic rather than static.

Decomposition-based fragmentation is consistent with the idea that problems can be solved and improvements made piece by piece. This means that once a critical part or function has been found – for instance in a manufacturing process or in a service chain – the critical part or function can simply be replaced by something better. The substitution myth described in Chapter 4 is the common assumption that something, a component or function, can be value neutral in the sense that its introduction into a system has only intended and no unintended effects. The substitution myth is, however, invalid since any change to a system necessarily disturbs an often delicate balance between resources and demands (Hollnagel & Woods, 2005, p. 101).

Fragmentation in time

The unspoken assumption behind PDSA is that the surroundings are stable while the change takes effect. This is another way of saying that what happens in the surroundings is deterministic. If it is stable, then it is, of course, deterministic because nothing happens unless something is done. It is also deterministic in the sense that the "plan" can precisely predict the consequence (which is the subject of "check" or "study") and hence that the expected outcome will be achieved.

It is important for PDSA to assume that the surroundings are stable since it otherwise would be impossible to achieve the desired change by a series of discrete steps, one following the other in a pre-defined sequence. Even if the steps are repeated as in the cycles, they are still fragmented or separated in time. It has even been argued that a "hygienic" separation of the steps in time is necessary since it otherwise would be difficult to determine the effects of a change.

> The virtue of the P. D. C. A. is not in Planning, Doing, Checking or Acting but in the separation of Planning from Doing, Doing from Checking and Checking from Action. It is a methodology that ensures that a change to a process such as one improvement is isolated from the following change.
>
> (Berengueres, 2007, p. 72)

The implication is that nothing happens except the steps that are part of making the change – i.e., the system undergoing change remains stable apart from the intended outcomes of the change. This is, of course, particularly important for the "do" step. If something else could happen while the planned intervention took place, then it might be difficult to conclude that the outcome was indeed the result of the intervention.

The assumption that the system undergoing change remains passive during the change was presumably quite reasonable at the time the Shewhart cycle was proposed and for the contexts it considered. But as already noted, while the assumption may hold true for technical systems – at least the technical systems of yesteryear – it does not hold true for social systems or for socio-technical systems. Lewin's spiral of steps indirectly recognised that through the need to restructure the overall plan by including the freeze phase after the change has been implemented. The OODA loop is even more explicit in this respect because it is intended for situations where the conditions change rapidly. The solution is to make quick decisions – so quick that the opponent does not have much possibility to do anything in between. Change management of socio-technical systems does not usually include an adversary and therefore does not need to put the same premium on speed. By contrast, we know that social systems are never stable but always change, mainly because they are open systems in partly unknown surroundings (quite apart from the entropy aspect). We shall return to this issue in Chapter 6.

The pedigree of decomposition is, of course, indisputable because it is the very foundation of science and the scientific method to do things by steps. Indeed, it is inevitable that we have to break sequences into parts when we think of something, both ahead of time when it is planned and afterwards when events are analysed. But the order of events one after another can also be seen as an artefact of one-dimensional time. Even if two things happen simultaneously it is always possible to increase the resolution to show that one actually came before – or after – the other. Yet the simple ordering in a sequence should not be mistaken for a cause-effect relation, although it often is. In reality, things happen one step after the other simply because we – as individuals – are unable to do two things simultaneously. Due to limited attention it is also much easier to control something if it is distributed in time than if it happens at the same time. But the reason why this works in science and in dealing with nature – at least in the classical sense – is also the reason why it does not work for change management in complex socio-technical systems. They are simply not stable for long enough.

Signal versus noise

Another assumption underlying the *ceteris paribus* principle and the various proposals for change management methods is that the planned interventions – the actions taken – are the causes of or reasons for the observed outcomes. Returning to the concepts of signal and noise discussed in Chapter 2, the planned changes are the signal, and the signal is supposed to be stronger than the noise. (*Noise* is a general term for the unwanted and usually uncontrollable changes that may interfere with, distort, or weaken a signal. Noise can also be seen as that part of the output of a system that is not determined by its inputs.) Indeed, the *ceteris paribus* principle implies that there is no noise worth speaking of. The question is, however, whether change management can take for granted that there are no extraneous "inputs" that may have as much effect – or even a larger effect – as the "signal."

Organisations today, as socio-technical systems, are always undergoing change due to internal and external forces, many of which are unknown and most of which are unpredictable. Agile organisations are able to cope with this in various ways as described – e.g., by resilience engineering (Hollnagel et al., 2011). Any planned changes happen on top of the constant adjustments and must create a synergy with these to have any chance of succeeding. This can also be described in terms of requisite variety and the law of requisite variety. To manage a change and control a system, the controller, which in this case can be seen as the plans for change management, must know what "naturally" happens in the system while the intervention ("do") takes place. But what "naturally" happens is in this case the noise or the "natural" variety about which we usually know very little.

All change management approaches appear to assume that the planned intervention, the signal, is so strong that any other influences or noise can be disregarded, at least for the time it takes to carry out an intervention. This can also be seen as a kind of fragmentation because it, so to speak, isolates the intervention from the dynamics of the rest of the system. Yet this is a risky assumption to make (to be discussed more in Chapter 6).

To invoke the voyage metaphor once more, classical change management such as PDSA can be likened to sailing in a motor-driven vessel. In a motorboat it is possible to set the course and speed even if it obviously is necessary to check the position every so often. Real change is unfortunately more like sailing by the wind. Here the ability to follow the intended course depends on the direction and force of the wind, on unpredictable lulls and squalls, on waves and currents, and so on. In consequence of that it is not always possible to move or sail as one would like independently of what is happening in the surrounding area. Sailing by wind, whether racing or for leisure, necessitates the use of tacking in a larger pattern that in the end ensures that the destination is reached, but not always from the direction that was intended or with the speed that was expected.

Change management for socio-technical systems needs an approach that carefully times small interventions to produce a synergy between the uncontrolled variability of the surroundings and the compensating responses. Actions never take place in a sterile or neutral world, but must always compensate for changes that occur independently of the planned actions at the same time as these try to move the system in the intended direction. Since such compensating actions have to take place anyway, it makes sense to use them as a basis for and in synergy with the intended changes that, e.g., the PDSA contemplates.

The PDSA cycle assumes that the system where the changes are being introduced is a closed system. It is furthermore assumed to be passive in the sense that it responds (as expected) to interventions but otherwise keeps functioning as it was supposed and designed to do. In this case the signal can therefore be assumed to be quite strong and to dominate any possible noise. If there is an issue of create-destroy in PDSA, it is about finding the right signal to begin with, not of substituting it during the change.

In the spiral of steps, the target system is a social system, which is dynamic and active rather than passive. The signal is based on the overall plan, but the overall plan is continuously being assessed and revised. It is, however, not being destroyed and (re)created. Instead, the repeated assessment can be seen as a kind of truth maintenance (in the AI sense of the term), i.e., making sure that the assumptions underlying the overall plan still are fulfilled, that the noise remains noise rather than gaining so much strength that it becomes an uncontrollable extraneous signal. If that happens, it should hopefully be caught in the reassessment of the overall plan. No such assessment – or "mindfulness" – is present in the PDSA cycle.

Finally, the OODA loop explicitly deals with a dynamic and changing world. The difference from the two other main approaches is that the changes are rapid and for the near term (or even the immediate near term), rather than for the longer term, as in the other cases. The OODA loop also assumes that the changes in the surroundings are supraliminal and that they are strong signals rather than weak signals, which means that they can easily be detected and recognised. The span or scope of the plan, which must be the outcome of the decision, is also quite short. It is therefore reasonable to assume that the decision, or the outcome of the decision, in fact is the signal and that there is relatively little noise during the act. This also seems to be reasonable because the approach looks at adversarial situations rather than long-term developments.

The problem in all three cases is that the approaches are used for situations and conditions that in many ways differ from those for which they were developed. This means that it cannot be taken for granted that they will work today as well as they did originally. Chapter 6 will look at how it may be possible to overcome the three types of fragmentation as the starting point for developing an approach to synetic change management.

6 Synetic change management

Introduction

Chapter 4 argued that it is necessary to constantly pay attention to how an organisation performs and functions, partly because of the continuous increase in entropy and disorder and partly to guard against the consequences of our fragmented understanding. Making and managing changes is never as simple as setting a target or goal and then identifying the interventions or activities that will bring about the outcomes. In the world of today it is far from easy, and sometimes almost impossible, to understand how an organisation functions, to say nothing of what goes on in its surroundings. It is not enough to identify and describe the target and deduce what the "missing middle" is and what the activities or interventions are that will take us from the current position to the target. Making and managing changes is not the rational exercise that has been implied by ambitious intellectual frameworks, from the general problem solver to contemporary theories of change. It is, of course, still useful to think in terms of goals and means, but the path between the present state and the target can rarely be plotted in advance and then followed as a simple route map. The main reason is that internal and external conditions keep changing so that by the time a complete plan or strategy has been developed, the assumptions on which it was based – the premises about system states, resources, constraints, etc. – may no longer be fully valid. Neither can data gathering be completed at leisure before a change; it must be continued during the change, as emphasised already by Lewin's spiral of steps. The critical issues here are the correctness or validity of the internal model, of the assumptions that are made about what will happen in response to a given intervention, and also why a specific intervention is assumed to be effective in bringing about the desired results.

Chapter 5 argued that change management is fragmented in at least three different ways. It is fragmented in terms of foci, as illustrated by the four issues referred to throughout this book – productivity, quality, safety, and reliability. The four issues, and sometimes others as well, are pursued separately with their own sets of models and methods and typically assigned to different parts of an organisation. Change management is also fragmented in terms of scope because it looks at delimited activities, subsets of organisational structures or functions, or at the performance of specific roles, responsibilities, and professions. Finally, change management is fragmented in terms of time in the sense that the changes are assumed to have a well-defined duration, both with regard to when they begin and, more crucially, with regard to when they end, and particularly in terms of when the consequences or outcomes will have been firmly and permanently established.

This chapter will discuss each type of fragmentation and propose ways in which they can be overcome. Due to the very nature of the problems of fragmented change management it is impossible to provide simple solutions that will be comfortable to everyone. There are unfortunately only uncomfortable solutions, but the long-term benefits of spending the extra resources and energy/effort they require will more than outweigh the short-term gains and convenience that follow from remaining within the comfort zone. People are attracted to simple or even monolithic solutions because they fit well with the human preference to focus on one thing at a time. Due to the reasons for psychological fragmentation we are probably predisposed to solve problems one by one rather than to see them together, to pursue Depth-Before-Breadth rather than the other way around. We claim it is rational because that which is present is certain, while that which is yet to happen is potentially uncertain. The alternative, Breadth-Before-Depth, would require more thoroughness – hence the spending of seemingly needless effort before something is done – and possibly also lead to a slight delay in action.

When something is recognised as an anomaly the intuitive response is to try to resolve it as soon as possible, almost as an instinctive reaction. The philosopher Friedrich Nietzsche long ago wrote that "to trace something unfamiliar back to something familiar is at once a relief, a comfort and a satisfaction, while it also produces a feeling of power. The unfamiliar involves danger, anxiety and care – the fundamental instinct is to get rid of these painful circumstances. First principle – any explanation is better than none at all" (Nietzsche, 2007; org. 1895). The urge to quickly relieve uncertainty is universally irresistible. People always try to solve the most pressing or recent problem – that which causes the most concern at a given time and therefore attracts attention – and many times optimistically apply readily available solutions without ensuring that they actually are going to work. This is done despite the experience that, in nearly all cases, it often is worthwhile to spend a little time understanding the breadth of the problem and seeing if other issues may be involved.

Dealing with the fragmentation of foci

Fragmentation of foci means that each issue is addressed by itself without considering how it may affect, or be affected by, other issues. A simple example is that safety is dealt with by the safety department while quality is dealt with by the quality department, or, as we have seen in Chapter 5, that there are suggestions for a safety culture, a quality culture, a production culture, and even a reliability culture. In fact, there is no shortage of specialised cultures that have been invoked as possible causes when the problem was diagnosed as the lack of a specific culture or as possible solutions when it was believed that the improvement of a specific culture would solve the problem. But it is clearly not sensible to willy-nilly postulate the existence of different cultures. At the very least they must bear some relation to each other since they must co-exist in the same organisation and be consolidated in the minds of people.

The earlier chapters have explained the reasons for the fragmentation of issues. It is, of course, impossible to reverse or negate the historical reasons for the fragmentation. Given how Western industrialised societies have developed, it probably could not have happened any other way. Neither can the psychological reasons for fragmentation be negated or nullified, except perhaps in unrealistic fantasies of universal AI taking over. But even if the reasons for the fragmentation cannot magically be removed, we can do something to overcome or compensate for the consequences, provided we are aware of them. This is what synesis is about.

As described in Chapter 2, the first issue is productivity, which in retrospect can be seen as an attempt to eliminate waste. Productivity has always been a concern for the efficiency of human work – even long before the Industrial Revolution. The earliest version of that may be found in armies, such as the Roman legions, where collective activities had to be effective in war as well as in peace. Taylor's concern was that people could do more and produce more than they thought it necessary to do, and that this would be to the benefit of the company as well as the employees. The problem was therefore in the observed waste of manpower –not because of deficient planning or organising but because people are usually local rather than global optimisers.

Signal and noise

The productivity problems Taylor tried to solve can be seen as due to the noise that arose from established working practices. When these problems had been dealt with and productivity had been improved to the point that it was acceptably high and stable, the influences of other sources of noise on productivity became noticeable. Taylor was concerned with manual labour – such as shovelling in the unloading of railroad cars full of ore, lifting and carrying in the moving of crude iron (iron pigs) at steel mills, manual inspection of bearing balls, and so on. Quality was therefore not the problem it later became in the manufacturing of consumer products. Here variability in quality had consequences for the market value of the products and hence indirectly for productivity. As Deming later argued, improving quality will not only reduce expenses but will also increase productivity and market share. Similarly, variability due to a lack of safety is also a problem because it can lead to disruptions or failure of productivity.

   When both quality and safety had been brought under (reasonable) control, the next issue, or source of noise, was the lack of stability or dependability of the processes, which was expressed in terms of the reliability of the parts, components, or functions that constituted the "machinery" of production. Some part of that had been taken care of by human factors engineering and the Men-Are-Better-At/Machines-Are-Better-At trade-off described in Chapter 3, although this happened in the pursuit of productivity and safety rather than in the pursuit of reliability. But the increasing use of ever more complicated technology – in particular computing machinery and later on information technology – introduced other concerns as described in Chapter 2. In addition to problems related to the reliability of the technology itself, the increasing intractability of the systems and processes that were taken into use made it necessary to consider two other sources of noise, one associated with the reliability of human performance and the other with the reliability of organisations.

   Productivity without quality is, however, of little use, just as quality without productivity is of little use. The same goes for the relation to safety and reliability. One way of overcoming the effects of fragmentation of foci is therefore to explore how the different issues mutually depend on each other. To do so it is useful to think of them as functions or activities, as something that is being done or takes place, rather than as states or conditions, as something that just is. In other words, it is useful to think of them as dynamic rather than static and hence to use verbs or verb phrases rather than nouns. Productivity would in this way become the act of producing or just "to produce." Quality would become the act of ensuring that the measured sample value falls between the upper and lower control limits or just "to ensure quality." Safety would become the act of reducing the number of unwanted outcomes to an affordable level (corresponding to a Safety-I perspective) or the act of ensuring that as much as possible goes well (corresponding to a Safety-II perspective) or just "to work safely." Finally, reliability would become the act of ensuring the presence and availability of all system-required functions whenever needed or just "to ensure reliability."

Enlarged view of work

A century ago Taylor, Shewhart, and Heinrich could allow themselves to focus on the specific work situation without thinking much about what happened in the larger system, to the extent that the term had any practical meaning. (The concept of a system, as in general systems theory, was not commonly used before the late 1940s.) This has left a legacy of models and methods that, although adequate when they were introduced, today are quite insufficient. A workplace, be it in production or services, can no longer be described simply in terms of how the parts – humans and machines – are organised and coordinated, not even if the organisational context is added. Organisations today must be understood as socio-technical systems or even as complex socio-technical systems. The socio-technical system concept was introduced in the mid-1960s (Emery & Trist, 1965) to focus on the relationships between workers and technology. But it was soon extended to account for how the conditions for successful organisational performance – and conversely for unsuccessful performance – depended on the interplay between social and technical factors. The concept has become ever more important due to rampant technological and societal developments that have dramatically changed the nature of work and how it is organised and managed.

To understand how work is done today it is necessary to enlarge the traditional focus of the single worker and his/her tools along three dimensions (cf. Figure 6.1). A "vertical" enlargement is needed to cover the entire system, from the supporting technology to the organisation, the latter famously referred to as the blunt end. One "horizontal" extension is needed to enlarge the scope of work to cover a larger part of the life cycle, from the design of equipment and infrastructures at one end to their maintenance and eventual cessation at the other. Finally, a second "horizontal" enlargement is needed because ongoing activities depend on what went before (upstream) and have consequences for what follows (downstream). An example from general production is the way in which "just in

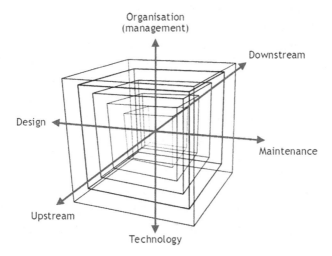

Figure 6.1 Enlarged view of work.

time" (JIT) is used to eliminate the need for inventories (raw materials, spare parts, etc.) but at the cost of becoming dependent on suppliers that may be in a different country or even on a different continent. Another example could be the gate-to-gate concept in aviation or the continual improvement concept that is Kaizen.

Because of these developments change management must today address systems that are larger and more intractable than the systems of yesteryear. Because there are many more details to consider, because some modes of operation may be incompletely known, because of tight couplings among functions, and because systems may change faster than they can be described, the net result is that many systems are underspecified or intractable, as discussed in Chapter 1. In these cases it is clearly not possible to prescribe tasks and actions in every detail. The conundrum is that due to the psychological fragmentation (again) we are unable to anticipate the full range of consequences of new technology or indeed of the attempts to improve productivity, safety, and quality that the increasing complexity forced us to make (Wright, 2004). Against our better knowledge we have developed and continue to develop and deploy systems and technologies that we do not completely understand and control but that we are unwilling – and perhaps also unable – to do without (cf. Figure 1.3). The incomplete understanding and control of such systems and processes gradually create an entanglement that challenges any change management philosophy.

On models (foci)

Synetic change management requires that we are able to understand and describe how multiple priorities or concerns are coupled and how they need each other to form or constitute a whole. Such an understanding is usually called a model, but as a model it must be more than a diagram with lines or arrows between components. The rendering shown in Figure 6.2 is therefore not a model, not even if the components are given more sophisticated shapes and colours. The general purpose of a model is to represent a selected set of characteristics or interdependencies of something, which we can call the target or object system.

> The basic defining characteristics of all models is the representation of some aspects of the world by a more abstract system. In applying a model, the investigator identifies objects and relations in the world with some elements and relations in the formal system.
>
> (Coombs, Dawes, & Tversky, 1970, p. 2)

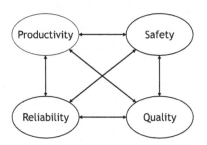

Figure 6.2 A diagram that is not a model.

The choice of characteristics is to a large extent defined by the purpose of the model or the formal system. A model of the brain could, for instance, be something that is similar in appearance to the brain (once it has been removed from the skull). In this case the nature of the material is more important than the functional aspects. The model could also represent something that functions in the same way as the object system in order to improve the understanding of that. In this case the functioning is important, whereas the material is not.

The diagram in Figure 6.2 is not a model because the arrows or connectors do not explain or describe how, e.g., a change in productivity will affect quality, how that in turn will affect safety, and so on. Since the diagram represents neither the structure nor the functioning of what we are trying to understand it is impossible from this kind of "model" to deduce anything about how the "system" functions or to get any help in deciding which changes will bring about which outcomes. In other words, it cannot be used to reason about the system it is supposed to represent. This is a common problem seen with simple block diagrams and flow charts.

In the case of change management a model should serve to capture or represent the internal ways of working of the system or the organisation that is to be managed. The model should help us to understand the internal "mechanisms" as a basis for choosing the interventions or activities that will lead to the desired outcomes. In the terms of the voyage metaphor, the model should enable us to steer the "vessel" or the organisation in the right direction – and at the proper "speed" – by making clear what needs to be done to move from the current position towards the target. One essential requirement is therefore that both the elements of the model and the relations between them are meaningful. Most models, however, fail rather miserably in this respect. A flow chart or rendering of the sequence of steps in an approach – for instance the PDSA in Figure 5.2 – is not a model since the relations between the four steps, P, D, S, and A, are undefined. It is, unfortunately, all too easy to create a flow chart and to link the various elements by lines and arrows. But unless the links are given a meaning, the result remains a graphical rendering and not a model.

A first step to providing a model that describes the mutual relations of the four issues could be a causal loop model like the one shown in Figure 4.3. This would require that the issues are defined in the same way as states. An example is shown in Figure 6.3. Here "productivity" means "the state of being productive" (and so on).

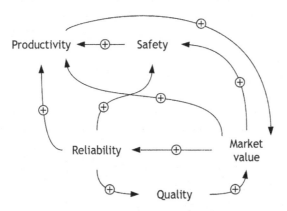

Figure 6.3 The four issues as a causal diagram (causal loop model).

To make the model meaningful, it has been necessary to introduce a fifth entity called "market value," which represents the revenue or, more properly, the profit. The market value provides the resources needed for productivity, safety, quality, and reliability. This simple model suggests that market value increases if productivity and quality increase; conversely, it decreases if either of these decreases. Although the model in Figure 6.3 is extremely simple it does indicate that improved safety and reliability will lead directly to improved productivity, while improved quality leads indirectly to improved productivity. This may obviously be disputed, but the importance of a model like this is that it provides a way to express the assumptions – or understanding – of how the issues are related and how they depend on each other. Notice also that all the relations shown in this model are directly proportional. This means that a positive development, once started, will continue to feed itself; but it also means that a negative development – for instance reduced reliability – will likewise feed itself and lead to a gradually worsening situation. While this is unrealistically simple, the use of causal loops nevertheless makes it possible to think more concretely about how an organisation could be managed and what should be done to keep it on the right track.

The logical next step would be to represent the issues as functions rather than as states, as suggested above. This would make sense in relation to change management since the purpose of this is to change how something takes place in order to produce a different outcome. What is changed, therefore, are ways in which something is done – namely, activities or, generically, functions.

Developing details of a functional model

One approach to developing a description of the issues as functions is the functional resonance analysis method (FRAM; Hollnagel, 2012). FRAM is a method of producing a functional model of how something is being done – in this case how an organisation works – by interpreting the four issues as representing four functions. The basic principle of FRAM is that a function is described by what it does rather than by what it is; it is described by what the outcome or output is and by what is needed to produce the output. The description of a function therefore typically includes its inputs and outputs, but it may also include four other aspects called preconditions, resources, control, and time. The preconditions represent something that must be true or verified before a function can begin; the resources represent that which is needed or consumed while a function is carried out; the controls represent that which supervises or regulates a function while it is carried out; and time represents the various ways in which time and temporal conditions can affect how a function is carried out.

One way to begin would be to ask what the outputs are from productivity, quality, safety, and reliability. In other words, what is the result of the associated functions, what do they bring about, or what would be missing if they did not perform as they should? Following that we could ask what the inputs to productivity, quality, safety, reliability would be. In FRAM, the inputs represent inputs in the traditional sense – i.e., that which is being processed and changed by a function – but also the conditions that trigger or start a function. The same procedure could be carried out for the aspects called preconditions, resources, control, and time. By asking such questions it will gradually become clear what the four issues actually entail as functions or activities and therefore how they may be mutually dependent. Without going into all the details as to what this may bring to light, a possible and deliberately simplified rendering is shown in Figure 6.4.

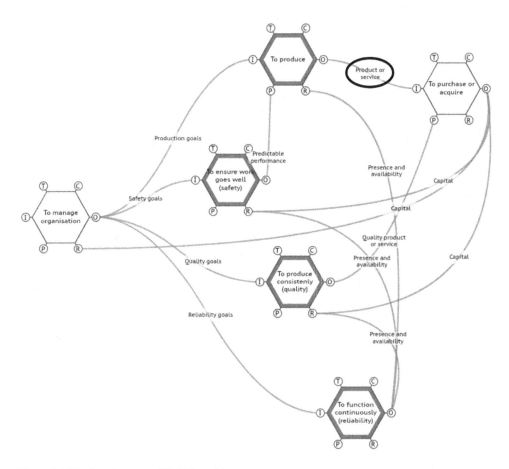

Figure 6.4 The four issues as a FRAM model.

There are a number of important differences between Figures 6.3 and 6.4. First of all, the model units or parts shown by the hexagrams are now functions or activities rather than states or objects. The entity that in Figure 6.3 was named "market value" is called "to purchase or acquire" in Figure 6.4. The output of the production is a product or service with market value and hence something that customers would want to purchase or acquire (the black oval in Figure 6.4 only serves to emphasise the output from the function "to produce" and is not a semantic element as such). Figure 6.4 also includes an additional function called "to manage organisation." This obviously needs to be described in far more detail – and hence expanded into a number of additional functions – since it is the management of the organisation and the changes the model serves to describe. The simple FRAM model in Figure 6.4 illustrates how it is possible to develop a description of the relations between the issues and how this can help to provide an understanding of how an organisation functions. The details are few and rough and will easily give rise to questions and arguments. But that is also precisely the purpose of a model. By describing the functions and their relations in a meaningful way it becomes possible to question whether the model is suitable or correct. That is not possible for a "model" such as that shown in Figure 6.2.

Overcoming fragmentation of foci

To overcome the fragmentation of foci it is necessary to find a way to see and treat the problems and issues together rather than separately. The examples shown in Figures 6.3 and 6.4 hopefully provide an idea about how this can be done. The world will remain unmanageable as long as we keep thinking in silos and specialities instead of the system, and systems, as a whole. Synesis shows how this can be done.

While we can begin to overcome the historically fragmented foci, it is less easy to overcome the psychological reasons for fragmentation. This is because it is natural and nearly unavoidable for our minds – or brains – to understand, reason about, and analyse the world we are faced with in a piecemeal fashion. Doing so defines our comfort zone or natural way of perceiving and understanding the world – what we do, what we experience, and what we are confronted with. To overcome the fragmentation that is due to psychological factors it is necessary to leave the comfort zone. (By contrast, many have tried their best to reduce cognitive workload to remain within the comfort zone. This solution unfortunately defies itself since it does not solve the problems but only moves them from the foreground – the interface and interaction – to the background.) To leave the comfort zone requires a deliberate effort and is probably impossible to do constantly – although it may have become necessary. The irony is that the consequences of psychological fragmentation – namely, the increasing entanglement of the world – make it necessary to overcome that which caused it. At the same time the very nature of or reasons for the fragmentation make it impossible to overcome, barring some revolutionary but unlikely improvement in how the human mind works.

Dealing with the fragmentation of scope

The second problem to overcome is the fragmentation of scope. This can also be described as the whole-parts problem or the system-surroundings problem. In practice it means that we chose to deal with a subset or part of the system rather than the system as a whole – which admittedly is an ill-defined concept. In combination with the fragmentation of foci it means that we look at a part of the system from only a single point of view. The pragmatic justification for accepting the fragmentation of scope is that it will be sufficient to deal with the subset or part of the organisation where the problems are located (such as a particular department or organisational unit), a particular concern (such as communication), a particular person or role, and so on. The practical approach is temporarily to define a narrower boundary, which includes the problem in its local context but considers everything else as part of the surroundings. This artificially limits the scope of the changes that are considered, although usually without fully understanding whether this is reasonable.

Dynamics

Whenever a change or an improvement is planned, it is tacitly assumed that we know what will happen and, in particular, that nothing unexpected will happen that may jeopardise the outcome of the intended changes either inside the (temporary) boundary or in the surroundings. The latter is, of course, a crucial condition that corresponds to the *ceteris paribus* assumption described in Chapter 4. This may have been a reasonable assumption to make in industries a century ago, when the rate of change and innovation in the work environment was relatively slow. Furthermore, when a change was introduced it was reasonable to assume that nothing else happened that would affect the work process and the working conditions while the change was rolled out.

Yet the same assumption cannot reasonably be made today. On the contrary, changes happen all the time, both inside a system and, more importantly, in its surroundings. It may be possible to narrow or limit the scope of the change so that the "local" system itself, as defined by the temporary boundary, can be assumed to be stable. But the limitation in scope means that more is included in the surroundings, which means it is more likely that changes will happen there. (It is like entropy, which cannot be eliminated but only distributed.) Since the changes that happen in the surroundings by and large are unpredictable, mostly uncontrolled, and quite possibly larger and with greater impact than the planned changes and improvements, the veracity of the *ceteris paribus* principle can no longer be taken for granted. The initial, and naive, response has been to try to exert greater control over the conditions, but to little avail.

When a change is started one important condition is that the system's performance is relatively stable or in a state of equilibrium (this does not mean that the state is acceptable; if that were the case, there would be no need of a change). Introducing a change will obviously affect and likely disturb or disrupt the equilibrium. But the hope is that the system, after a while, will come to rest in a new state of equilibrium that is different in the intended direction. How long this may take is not easy to predict; expect that it probably will take longer than usually assumed. The idealised development can be represented as shown in Figure 6.5.

Figure 6.5 shows the change as taking place in isolation, but the planned change is never the only thing that happens in an organisation. Even if a reduced scope excludes other changes from the planning, they will nevertheless happen outside the "local" system. The situation is therefore more as shown in Figure 6.6. The intended change (represented in bold) is just one of the many changes that happen throughout the system – some before, some overlapping in time, and some following. One problem is that no one knows how many changes happen at the same time. A second, and potentially more serious, problem is that no one knows how these multiple changes may affect each other, either directly or indirectly. Since the changes are beyond control, and since they are also unpredictable – or, even worse, often ignored – they may render the intended change ineffective. The wise thing to do is therefore to try to understand as well as possible what is likely to happen outside the chosen boundary and to find a way to create synergy with that.

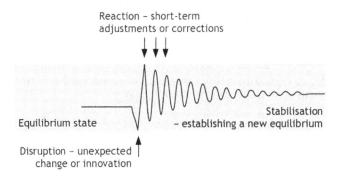

Figure 6.5 Equilibrium – disturbance – equilibrium.

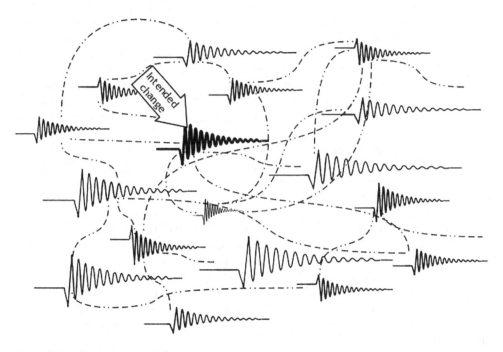

Figure 6.6 A change in a sea of changes.

Relative to the intended change or signal, the variability of the surroundings can be seen as a source of noise, where the noise sometimes may be stronger than the signal. Since simply strengthening or amplifying the signal (the intended change) is unlikely to be a solution, the alternative is to see whether it is possible to combine the signal and the noise, thereby using the latter to amplify the former. This is possible not only in principle but also in practice because the noise is far from random or stochastic. The surroundings are not a uniform grey mass, but rather an agglomeration of other systems where the definition of what the system is and what the surroundings are depends on the point of view. In an assembly of people, each person is from his or her position the "I," while the rest of the assembly are "the others." In the same way each system will consider the agglomeration of other systems as its surroundings. Since it is in the interest of each system to sustain itself, the system's variability will be purposeful from its own standpoint, although others initially may see it as random. Because it is purposeful it is in principle possible to anticipate what could happen, which means that the regularity of or patterns in the noise can be recognised as a "signal" that can be used for an intended purpose – as in the coracle analogy described in the following section. As long as the variability has some kind of recognisable order or regularity, it may serve as an organisational structure or scaffold for the change. This will be described further in Chapter 7.

The coracle

As an illustration of how the noise from the surroundings can be overcome, consider the story of how Jim Hawkins struggled with manoeuvring the coracle, as told by Robert Louis Stevenson in the novel *Treasure Island*. Jim Hawkins, having sneaked out from the stockade, intended to use the coracle to row out to the Hispaniola and cut her adrift. The problem was that the coracle turned out to be very difficult to control and completely impossible to paddle in the intended direction. It was described as "like the first and the worst coracle ever made by man" (Stevenson, 1969; org. 1883, p. 139). In terms of manoeuvrability, it was "the most cross-grained, lop-sided craft to manage. Do as you pleased, she always made more leeway than anything else, and turning round and round was the manoeuvre she was best at. . . . Certainly I did not know her way. She turned in every direction but the one I was bound to go; the most part of the time we were broadside on" (Ibid., p. 141). Yet Jim also gradually realised that the coracle, when left to itself, would effortlessly ride the waves and "would but bounce a little, dance as if on springs, and subside on the other side into the trough as lightly as a bird" (Ibid., p. 147). But it did not take well to paddling.

> I began after a little to grow very bold, and sat up to try my skill at paddling. But even a small change in the disposition of the weight will produce violent changes in the behaviour of a coracle. And I had hardly moved before the boat, giving up at once her gentle dancing movement, ran straight down a slope of water so steep that it made me giddy, and stuck her nose, with a spout of spray, deep into the side of the next wave.
>
> (Ibid., p. 147)

Put differently, the strength of the signal, the paddling, could not overcome the influence from the noise, the coracle's idiosyncratic way of moving on and with the water and the waves. Having realised that all attempts to control the course of the coracle by paddling alone were futile, Jim at length realised what he could do.

> "Well, now," thought I to myself, "it is plain I must lie where I am, and not disturb the balance; but it is plain, also, that I can put the paddle over the side, and from time to time, in smooth places, give her a shove or two toward land." No sooner thought upon than done. There I lay on my elbows, in the most trying attitude, and every now and again gave a weak stroke or two to turn her head to shore.
>
> (Ibid., p. 150)

He nevertheless had to give up reaching his intended target, the Cape of the Woods, and instead ended by boarding the Hispaniola. And we all know what happened after that.

Signal and noise – again!

In change management the planned change is by definition the signal and everything else is considered noise that may distort or block the signal. The conventional solution is to control, eliminate, or disregard the noise as far as possible in the firm belief that the signal in the end will prevail.

Synetic change management instead starts by acknowledging that there always will be uncontrollable variability or noise. It is therefore essential to understand the nature of it and, in particular, to find its regularities. Although the variability of performance in a socio-technical system at first may appear to be random, it is actually not. The variability is due to the approximate adjustments that both people and organisations make as part of everyday functioning. Since these adjustments are purposive it is not too difficult to recognise a relatively small number of recognisable shortcuts or heuristics. Performance variability, in other words, is semi-regular or semi-orderly and therefore also partly predictable. There is a regularity in how people behave, including how they respond to the variability of other people and the surroundings. Without that no place of work would be able to function.

The aim should, however, not be to suppress or quell the noise to enhance the impact of the signal. The intended change is, of course, still the one that has to be implemented, but synesis recognises that the noise sometimes may be stronger than the intended signal and that simply trying to force your way through is not going to work. The obvious alternative is to use the variability or noise in a synergistic manner.

An analogy can be found in space missions, such as sending a rocket to the moon. The straightforward solution is, as in Jules Verne's book *From the Earth to the Moon*, simply to build an enormous cannon, a columbiad space gun, aim it at the moon – or where you think the moon is going to be when the projectile arrives – and then fire. In practice the solution is not to try to overcome the gravitational forces directly but instead to use them in a gravitational slingshot manoeuver where the relative movement and gravity of astronomical objects, such as other planets, is used to alter the path and speed of the spacecraft. This is possible because the gravitational forces from the astronomical objects are orderly and predictable. In the same way, an alternative to making a brute force change to an organisation is to make use of the events and changes that are going to take place anyway, whether we want them to or not. Planning to make a change should therefore start by trying to develop as good an understanding as possible of what is likely to happen in the organisation where the change/improvement is going to take place, including, of course, what could happen in the surroundings. We should, as Jim Hawkins did in the coracle, look for and find the regularity or patterns in the noise and then use that to achieve our aim. We need to work together with the "natural" forces rather than battle against them even if it does not immediately bring us where we want to go. To do that we need a good functional model of how the waves occur, which means that we need to determine what changes will happen outside the limited scope. This may be done by applying the same techniques and methods that are used to overcome the fragmentation of foci.

Overcoming fragmentation of scope

To overcome the fragmentation of scope it is necessary to recognise that it is impossible to make a clean cut or to define a clear boundary between the system and the surroundings, i.e., to assume that there is a well-defined and easily recognisable distinction between them. Furthermore, it is necessary to realise that making the boundary narrower or tighter, and thereby reducing the scope even further, does not work. It will not actually remove the variability or noise but only push it into the surroundings, which thereby become more difficult to understand.

There is, of course, nothing wrong with having big plans and having ambitious goals that go beyond the short term. Without big plans there can be no real development and no innovation, only local and temporary adjustments and hence a loss of initiative. But it is not advisable to rely on big steps as the only way to realise big plans. Big steps take a long time to implement, and it may be even longer before any results can be seen. When a change finally occurs it is difficult or impossible to know whether the results – or lack of same – are due to what was done or to something else that happened during the same time. Big steps also require big investments, and people tend to remain committed to big investments – a phenomenon known as escalation of commitment (Lofquist & Lines, 2017) – which may skew further progress.

The obvious alternative is to implement plans – big, medium, or small – through small steps. This means that a long-term goal is broken down into more short-term subgoals that can be reached more quickly. It takes time for effects (and influences) to propagate. If the steps are small or brief (i.e., of short duration), it becomes more reasonable to assume that the surroundings remain stable, neither affected by the changes nor affecting the steps. The same argument can be used for the surroundings because the definition of system/surroundings is reciprocal. The subgoals should be chosen such that the expected time to achieve them is shorter than the mean time between other – uncontrolled – changes. If that can be done it means that the surroundings can, in effect, be considered stable while the smaller change takes place. In the story of the coracle, Jim Hawkins would row only in smooth places where the paddling could actually make a change. The changes in the surroundings to look out for could be market developments, customer preferences, economic cycles or disturbances, political changes (e.g., elections), etc. In most cases people will have a pretty good understanding of where and when some kind of temporary stability may be likely or, conversely, when it is highly unlikely. And if they do not, they should try to find out.

By choosing step sizes that correspond to periods or intervals of relative stability in the surroundings there is good reason to expect that the observed changes will be a consequence or result of the chosen interventions rather than of something else. These observations can then be used to reassess the larger goal and to consider what the next small step should be. This is actually similar to Lewin's cycle of steps, except that we now have some kind of guidance to how large or long the cycles should be. Other factors that may determine the step size or duration are, of course, the nature of the process, the efforts and investments required, what is practically feasible and how it fits with whatever else an organisation does, etc.

A deliberate use of small steps is technically known as incrementalism, which is the method of working by adding to a project using many small incremental changes instead of a few large (and expensive) jumps. It implies that the steps in the process are sensible, which means they are small enough to be reasonably well understood and also short enough in duration that it is reasonable to assume that nothing else significant will happen during the time it takes to carry them out. Incrementalism is another way of describing "muddling through," the compromise between the rational actor model and bounded rationality developed by Lindblom. By contrast, "non-incremental policy proposals are therefore typically not only politically irrelevant but also unpredictable in their consequences" (Lindblom, 1959, p. 85).

Dealing with the fragmentation in time

The third type of fragmentation is with respect to time. A necessary step in planning a change is estimating its duration, both in terms of when it begins and, more crucially, in terms of when it ends. The latter is especially important because the purpose of the change is to ensure some specific consequences or outcomes. It is obviously assumed that they will have been firmly – and possibly permanently – established at some point in time. In today's business and production environments it is even more important that there is a point in time where the result of the changes can be measured so that a successful outcome can be declared. The determination of when that may happen should, as far as possible, be based on facts, but it is often not. While processes and activities undeniably are continuous rather than discrete it will in practice be necessary to divide a development into smaller sequences or "time windows" and to assume that everything of interest and importance happens within a time window. A major advantage is that this limits or reduces what it is necessary to pay attention to and thereby limits the resources and investments that have to be set aside. Yet even within that time window of development we tend to divide it into smaller, discrete steps. This is partly justified by the legacy of linear causality.

Linear causality

Making a change can be seen as a way of ruffling or disturbing an equilibrium (e.g. Figure 6.5) rather than as injecting a cause that has a clear, predictable effect. In Western philosophy (but also in other cultures) the concept of a cause-effect relation is the fundamental way of explaining why and how something happens. In his *Metaphysics* (Book v, Part 2) Aristotle defined cause as "that from which the change or the resting from change first begins; . . . in general the maker a cause of the thing made and the change-producing of the changing." In the 18th century the Scottish philosopher David Hume put forward eight rules by which to judge causes and effects, the first four of which are reflected by the following:

(1)   The cause and effect must be contiguous in space and time.
(2)   The cause must be prior to the effect.
(3)   There must be a constant union betwixt the cause and effect. 'Tis chiefly this quality, that constitutes the relation.
(4)   The same cause always produces the same effect, and the same effect never arises but from the same cause.

(Hume, 1985; org. 1739–40, p. 223)

These rules were undoubtedly reasonable in the 18th century, at least in a treatise on the human mind, but they are clearly not reasonable in the world of today. They can nevertheless easily be recognised, not least in how we deal with adverse events and other forms of unwanted outcomes in relation to production, quality, and safety.

The very idea of cause-effect relations serves to break a continuous flow – or flows – of events into parts, segments, or fragments. Fragmentation in time assumes linearity and tractability (that systems are tractable, as discussed in Chapter 1). If these assumptions are correct, then the steps can be considered individually within their designated time windows. But if the assumptions are wrong, this cannot be done. Although it may be convenient we cannot start by assuming that actions can take place in a sequence, let alone an immutable sequence. Actions are intended to bring about some changes, and we must therefore begin by understanding how actions are dependent on each other, both in the sense that outcomes are not in a 1:1 relation to individual actions and in the sense that actions – or functions – may have preconditions, require resources, etc. More importantly, we need to understand when – and where – we can expect to see the results of interventions.

The importance of Hume's thinking in this context is the contiguity of cause and effect in space and time described by the first rule since that effectively establishes the justification for the fragmentation in time. If causes and effects are contiguous in time, then it is reasonable to consider changes within a limited window of time. This becomes even more reasonable if linear causality is also assumed since that makes it possible both to predict the outcomes of an intervention and to determine the initial or root cause of a change. Unfortunately for this argument, the contiguity in time no longer holds. (In relation to the fragmentation of scope, the first rule is also important because it emphasises that cause and effect must be close together in space. Chaos theory has convincingly demonstrated that this is not always a valid rule.)

One problem for the traditional way of cause-effect thinking is the concept of emergence put forward by the English philosopher George Henry Lewes (1817–1878). He argued for a distinction between resultant and emergent effects and wrote:

> Thus, although each effect is the result of its components, the product of its factors, we cannot always trace the steps of the process, so as to see in the product the mode of operation of each factor. In the latter case, I propose to call the effect an emergent. It arises out of the combined agencies, but in a form which does not display the agents in action.
>
> (Lewes, 1875, p. 368)

The concept or idea of emergence corresponds to the notion that the whole is more than the sum of its parts. In relation to change management it simply means that it is not always possible to determine or find the causes of an observed effect, and conversely that it is not always possible to predict the effects of a set of causes. This obviously has consequences for the fragmentation in time, for choosing to work within a given time window, since we cannot be sure that emergent effects will occur within that time window. If there are no discernable effects, there is no feedback. And without feedback it is impossible to manage the changes.

There is, unfortunately, also a second, and considerably larger, problem – namely, Erwin Schrödinger's concept of entanglement. This is a quantum mechanical phenomenon according to which the quantum states of two or more objects must be described with reference to each other even though the individual objects may be spatially separated. In quantum theory it means that there can be correlations between observable properties of systems even though this violates the speed limit on the transmission of information implied by the theory of relativity (Albert Einstein famously referred to this as "spooky action at a distance"). Entanglement is not an actual phenomenon in the macroscopic world, but the concept provides a way to characterise socio-technical systems that have developed beyond intractability. In relation to change management, entanglement is used in an analogous rather than a direct sense to describe how there can be effects that defy traditional thinking because they seem to occur out of nowhere ("spooky") and because they are either much faster or much slower than anyone expected. Both emergent effects and entanglement are congruous to Charles Perrow's (1984) ideas of complex systems and non-linear outcomes even though entanglement as such is a step beyond the complexity Perrow described. Altogether this means that the fragmentation of time should be treated very carefully and that limiting the consideration of changes to a small and well-defined time window is likely to lead to unexpected results/effects.

Things take time

In 1846 the English scientist Michael Faraday held a Friday Evening Discourse at the Royal Society in London. As part of this Discourse he asked his audience to consider the following thought experiment:

> What would happen if the Earth was suddenly dropped into place, at its proper distance from the Sun? How would the Sun "know" that the Earth was there? How would the Earth respond to the presence of the Sun?
>
> (Gribbin, 2003, p. 423)

Faraday used the thought experiment to support his idea of a field of forces, something that was far ahead of the thinking at the time. In the current context the thought experiment can be used to illustrate the time it takes for a change to have effect. Instead of suddenly putting the earth at its proper place in the solar system, consider what happens when an intervention, such as a change to how work should be done, is made in a system or organisation. The change as such (new rules, new equipment, new ways of working, new priorities) will immediately be subjected to the existing norms and habits of the people who do the work. (This is one reason why Lewin proposed that the first step of a successful change should be to unfreeze existing norms and habits.) They will receive and interpret the change according to their established routines and practices, just as the earth immediately would "feel" the influence from the sun and the other planets. But it will take some time before other people and other parts of the organisation become aware of or are affected by the change, just as it would take some time before the sun would "know" that the earth had suddenly appeared. In Faraday's thought experiment we can say precisely how long it would take since it is determined by the speed of gravity (although this was not known during Faraday's time). But in the case of making an intervention in an organisation, no one knows at what "speed" the effects will propagate. Some effects, such as "news" or rumours, can spread very quickly and through channels and links that are mostly unknown. Other effects, such as the apocryphal changes in culture, will take a long time, although no one has any precise idea of just how long. And unlike the speed of gravity and the speed of light, there is not one common speed of propagation of effects following organisational changes, but rather many, although their number and nature are anybody's guess.

A recent illustration of how little attention is paid to understanding the likely duration of a change is provided by the ambitious attempt to improve education in primary schools in Denmark. In 2014 a school reform was implemented that should make students more proficient in Danish and mathematics. The six basic elements of the reform were an increased number of lessons, supportive teaching, physical movement and exercise, open school, homework help, and clarifying the role of educators. In 2019 the effects of the reform were analysed using responses from about 10,800 teachers, 2,300 educators, and 1,500 school leaders as well as register data from Statistics Denmark and from over 400,000 students. The conclusion was that students had not become more proficient in Danish and mathematics. In the slightly heated discussion of why the expected results could not be found, one of the political parties that had supported reform meekly pointed out that it might take five to fifteen years for a school reform like this to be fully implemented. No one, however, seems to have been concerned about that when the reform was launched; otherwise they would hopefully not have carried out the so-called final evaluation in 2019, less than five years after the start of the reform.

Minimising fragmentation in time

Whereas it is possible in principle to overcome the fragmentations in foci and in scope, it is not as easy to find a principle or approach that will overcome the fragmentation in time. The reason is not just that it is difficult to comprehend how an organisation develops over a long time scale (unless, of course, the view is limited to conspicuous historical trends that do not provide the details needed for change management). It is also that it is only possible to work with or contemplate events and changes of a limited duration. This is not the same as the limited attention span described in Chapter 2, which was concerned with the ability to concentrate on or pay attention to something without being distracted. It is, rather, that people find it difficult to imagine developments, such as making changes to a system, that do not have a well-defined end point. Yet to minimise the consequences of fragmentation in time it is necessary to accept that while the actions needed to bring about a change have a clear beginning and end, the same is not true for the effects. Although an effect obviously cannot be expected before the action has started (cf. Hume's second rule), it is far more difficult to say when it will happen and how long it will take before the final consequences can be seen. This is because there are always unknown or unrecognised couplings and dependencies within the system. It is nevertheless essential to be realistic about how long it will take for something – an intervention or a disturbance – to settle down. It is also necessary to accept that it can be quite difficult to know how long it will take for a change to have reached its end and hence to know when it is reasonable to look for or measure results. Finally, it is necessary to acknowledge the fact that little is known regarding how events are coupled, one reason being that organisations today are entangled rather than merely complex.

It may be useful to think of the time in terms of a window of opportunity for introducing a change. There will always be some point in time – commonly referred to as the earliest starting time – before which it is impossible to begin. Preparations may have to be made, such as the unfreezing described by Lewin; some activities may have to be completed, such as a production run or an ongoing project; and it may be necessary to wait until the effects of a previous intervention have subsided before a change can begin (assuming, of course, it is known when this occurred). Likewise, there is a time – known as the latest starting time – after which it is too late to begin. The window of opportunity to introduce a change is therefore at one end determined by the earliest starting time and at the other by the latest staring time. But while there is a reasonably clear understanding of when a change can begin, there is more uncertainty about when it will have ended. The interventions or actions that bring about the change will, of course, end at some point in time, but it is less certain when effects or consequences will show themselves and when they will have become stable. The former defines the time when it is reasonable to expect that the outcomes of the change will be observable or measurable. The latter defines the time when a new equilibrium will have been reached (cf. Figure 6.5). As the example provided by the Danish primary school reform shows, it is often, or even typically, assumed that the consequences will have been established by the time the project period – and funding – runs out (anyone who has tried to manage a project will be keenly aware of that). Yet there is little empirical evidence to help with actually determining when that is. The problem is simply that we rarely know in a factual sense how much time it takes before a change or intervention has been completed.

In addition to the problems related to the direct or intended effects there are also problems related to potential indirect or unintended effects. Due to the intractability or entanglement there may be indirect effects or unintended consequences that show themselves only a long time after the change "officially" has come to an end and that possibly may negate the intended effects. Although put forward more than seventy years ago, Lewin's suggestion of unfreeze and freeze therefore makes good sense even today. The purpose of the unfreeze should not only be to inform, prepare, motivate, and engage people who will be affected by the change or to examine the day-to-day activities and habits to clarify the premises for the change and ensure the correctness of the assumptions on which the change is based. The unfreeze should also serve to ensure that there are no hidden developments that could cause unwanted surprises. Similarly, when the change has come to an end, the purpose of the freeze is to make sure that a new state of equilibrium truly has been established before the improved conditions are consolidated. The change will intentionally have disturbed an equilibrium, and it may take a long time before work has settled down to a steady state again. Even if it can be difficult to estimate when this will occur it is necessary to do so or at least make a serious attempt at it. Otherwise, one of two things are very likely to happen: either there will be a lack of intended consequences at the time they were expected or there will be unexpected and unintended consequences for some time after.

To minimise the consequences of fragmentation in time it is necessary to be as realistic as possible about the timing and duration of the change – the interventions or activities as well as the consequences or effects. A change must, of course, always be of finite duration, but the expected duration should be based on a realistic assessment rather than tradition, convention, unfounded and optimistic guesses, political expediency, or wishful thinking. That a change must be planned and prepared to take place at a certain time and to be of a certain duration is unavoidable. The overriding concern in doing this is usually efficiency, i.e., to ensure that the intended and desired outcomes are achieved within time and budget. This can, however, not be done without a minimum of thoroughness in the preparations and planning, which in turn means that there must be a good understanding of how something is actually done (Work-as-Done). Changes should not be based on assumptions of what goes on (Work-as-Imagined) since these inevitably will be either underspecified or outright wrong. Thoroughness is required both to define the time window for the change and to select the right scope. In the first case it is essential to have a good understanding of the dynamics of the process and functions that are involved, and in the latter it is equally essential to understand the dynamics of what happens in the surroundings and outside the temporary boundary as well as any hidden dependencies. This is shown in Figure 6.7.

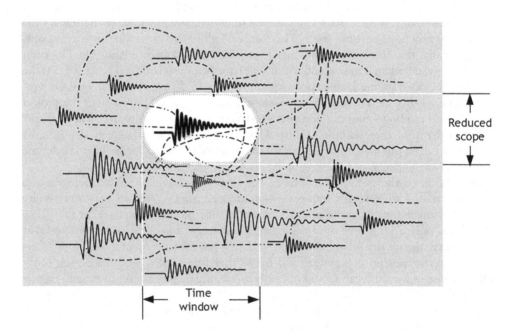

Figure 6.7 Requisite realism of change management.

The time window for the change should be defined based on a good understanding of the dynamics of the intervention as well as of whatever else is going on inside the temporary boundary ("reduced scope" in Figure 6.8) and the surroundings. Which "forces" or developments will the intervention set in motion and how will they develop? How do they depend on each other – or on other processes – and how may seemingly trivial outcomes aggregate to become non-trivial effects as described, e.g., by the concept of functional resonance (Hollnagel, 2012)?

The time window should also be defined relative to the chosen scope since the temporary boundary should be defined so that anything that may be the source of significant variability is inside the boundary (this will be discussed in more detail in Chapter 7). The scope must obviously be limited, but the definition of the boundary requires thoroughness rather than efficiency. It must be based on a good understanding of whatever else may happen while the change takes place rather than on convenience or habits.

The basic dynamics of making a change to a system

Figure 6.8 shows how causal loops can be used to describe the basic dynamics of making a change to a system. Start, for instance, in the lower left corner with the planning of the change, here called "Time to reflect and think ahead." The link to the following step, "Timely and appropriate actions," is marked with a "+" sign, which indicates a directly proportional relation. In other words, if more time and effort are spent planning and thinking ahead, it is more likely that the resulting actions will be timely and appropriate. Conversely, if less time and effort are spent on thinking ahead, it is less likely the ensuing actions will be timely and appropriate. The link between "Timely and appropriate actions" and "Anticipated developments of events" is marked in the same way, and the same reasoning applies here. The more timely and appropriate the actions are, the more likely it is that events will develop as anticipated.

The link between "Anticipated developments of events" and "Need to analyse feedback in detail" is marked with a "-" sign to indicate that it is assumed to be inversely proportional. This means that if the events develop as planned, there is less of a perceived need to analyse the results than if the developments do not go as planned. Conversely, if

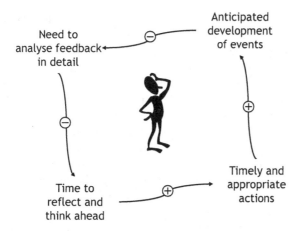

Figure 6.8 Change management as a causal loop.

the results or outcomes are unexpected, there is more need to analyse them to understand what happened. (This, however, does not mean that expected outcomes can be taken for granted and that one should not pay attention to them. It is important to be aware of the confirmation bias. But that is another story.) The final link between "Need to analyse feedback in detail" and "Time to reflect and think ahead" is also marked as inversely proportional, which simply means that if more time is needed to analyse the feedback, there will be less time available to plan the next step and hence it will be less likely that the actions will be appropriate (and so on).

The purpose of Figure 6.8 is to illustrate one way of capturing some of the basic dynamics of a process, but it is by no means a realistic model of change management (a model built by the FRAM would be more accurate but will not be attempted here). Neither should it be seen as some version of the cycles that are a central feature of the change models described in Chapter 5. The causal loop shown in Figure 6.8 emphasises the importance of being thorough, of spending sufficient time and effort during the planning and preparations of a change. If sufficient time is spent to reflect and think ahead, it is more likely the actions will be timely and appropriate. That being the case, it is also more likely that the outcomes will be as expected, which means that less time and effort are needed to analyse the feedback. This means that there (again) will be more time to think ahead and plan – hence a positive development that builds on itself. Conversely, if insufficient time is spent to reflect and think ahead, it is less likely the actions will be timely and appropriate and, hence, that the outcomes will be as expected. This means that more time and effort will be needed to analyse the feedback, thus leaving even less time to think ahead and plan – in other words, a negative development that will continue to make matters worse.

Strategy versus tactics

An important distinction in control, as well as in management, is between strategy and tactics. The concepts come from the military and the famous studies of war by Sun Tzu (c. 544–496 bc) and Carl von Clausewitz (1780–1831) but have been adapted to be used in many other contexts as well. A strategy can generally be defined as a plan of action, or a master plan, designed to achieve a long-term or overall aim (notice the similarity to Lewin's concept of the overall plan). Carl von Clausewitz writes:

> Strategy is the employment of the battle to gain the end of the war; it must therefore give an aim to the whole military action, which must be in accordance with the object of the war; in other words, strategy forms the plan of the war, and to the said aim it links the series of acts which are to lead to the same, that is to say, it makes the plans for the separate campaigns, and regulates the combats to be fought in each. As these are all things which to a great extent can only be determined on conjectures, some of which turn out incorrect, while a number of other arrangements pertaining to details cannot be made at all beforehand, it follows, as a matter of course, that strategy must go with the army to the field in order to arrange particulars on the spot, and to make the modifications in the general plan which incessantly become necessary in war. Strategy can therefore never take its hand from the work for a moment.
> (von Clausewitz, 1989; org. 1832, p. 177)

In modern management literature the term *strategy* first came into use in the 1960s. The meaning was, as in the military use, "the determination of the long-term goals of an enterprise as well as the courses of action and the allocation of resources necessary to reach these goals." Where a strategy is aimed at the overall objective, a tactic is a conceptual action aimed at the achievement of a goal that is part of the way to the target. This action can be implemented as one or more specific tasks.

The difference between strategy and tactics can be illustrated by using an example provided by Schützenberger (1954). Think of a person standing at the top of the hill who wishes to get to a house in the valley as quickly as possible. The way down, however, presents several obstacles, including boulders, streams, marshes, and edges, that make it impossible to take a direct path.

> An exhaustive, and final, solution of the problem would be given by taking a map of the district, dividing it into small areas, finding the time taken to cross each area individually, joining the areas into all possible chains between the top of the hill and the house, and then finding which chain gives the smallest total time for the journey. The path so selected is absolutely the best and has been selected by what I shall call the "strategy" of the problem. . . . Usually, of course, the traveller would not use so elaborate a method. A common method would be to make the selection in stages. He would first select a point about a hundred feet down to which he could get rapidly; then, arrived there, he would select another point a hundred feet lower still to which a rapid descent was possible, and so on till he reached the house. This method I shall call a simple tactic, as contrasted with the previous strategy. The tactic differs from the former in that the tactic does not take into account the whole of the situation, but proceeds according to a criterion of optimality that is applied locally, stage by stage.
>
> (Schützenberger, 1954, pp. 206–207)

In relation to change management tactics can be seen as corresponding to the principle of incrementalism described previously in the sense that changes are limited to smaller systems or subsystems that will remain stable vis-à-vis the surroundings while the change takes place. This represents an intentional and considered reduction of the scope in full realisation of the consequences rather than a reduction chosen merely for the sake of convenience.

A fourth fragmentation?

The unavoidable challenge of change management is the need to think of things together – unified rather than fragmented – even though this can be difficult to do for the reasons explained in Chapters 2 and 3. It is possible to overcome the historically based fragmentation by making a deliberate attempt to describe how the different issues are coupled and how they directly or indirectly depend on each other. This is essentially what this book has tried to do. But it is not possible to overcome or undo the fragmentation that is due to the way the human mind works. Instead, it is possible to compensate for them in the ways argued by this chapter.

The consequences of the fragmentation of foci can be compensated for by acknowledging that the issues depend on each other and by describing or modelling these dependencies in more detail using the techniques described in this chapter. The consequences of the fragmentation of scope can be limited by recognising the effects of how and where the boundary is drawn; by understanding the dynamics of the surroundings, not least whether there are signification sources of "noise"; and by choosing appropriately sized steps in accordance with the principles of incrementalism. Finally, the consequences of the fragmentation in time can be dealt with by being realistic about the timing and especially the duration of a change – the intervention itself as well as the effects or consequences.

It is a little ironic that the way to bring about the sorely needed unification of issues entails an analysis that points to the three types of fragmentation – foci, scope, and time – summarised earlier. Yet this fourth type of fragmentation – which perhaps should be called a meta-fragmentation – is required simply because it is humanly impossible, even with the best of intentions, to look at something as an undivided whole. This analysis is, however, only a first step that must never stand alone. It is crucial that it is followed by forging the synesis needed for the coherent management of current and future socio-technical systems.

7 A nexus of necessary knowledge

"What most experimenters take for granted before they begin their experiments is infinitely more interesting than any results to which their experiments lead."

(Norbert Wiener)

Introduction

A constant theme throughout this book is the need to understand what lies behind the challenges and opportunities that individuals and organisations face every day, not least when it comes to managing changes in organisations. Philosophers and great thinkers, as well as the rest of us, have seemingly forever tried to make sense of what happens in order to relieve anxiety, to be in control, to ensure that things develop as intended and needed, and to reduce unwelcome surprises to a minimum. The need to do so has not diminished as technologies and societies have developed more rapidly during the last fifty to seventy years; on the contrary, our inability to fully understand what happens has given rise to a never-ending series of supposed improvements that have often turned out to create more problems than they have solved.

There is probably – or perhaps definitely – no magical solution to this problem. But this should not stop us from making serious attempts to diagnose it more precisely and, especially, to try to find out which knowledge is lacking since that must be the starting point for research and development. Unlike the formal sciences, such as mathematics, geometry, and formal logic, or the natural sciences, such as biology, chemistry, and physics, there are no axioms, systematic taxonomies, or universally accepted classifications to be found in social and behavioural sciences. The best alternative is probably to organise the relevant knowledge – that which exists and that which is still being developed – relative to the practical problems that somehow must be solved. In the case of change management the experience accumulated over many years points to three topics that together can be seen as constituting a nexus of necessary knowledge.

The first topic is the knowledge that is necessary to understand systems, what they are and how they function. This must go beyond glib phrases such as a "systemic view" and try more precisely to characterise how an understanding of systems can be of help in change management. The second topic is knowledge about variability, the fact that humans and therefore socio-technical systems never perform as flawless machines even when misguided efforts are made to make them do so. This is the realisation that socio-technical systems, unlike machines, would be unable to function unless there was some variability. The third topic is knowledge about the regularities that can be seen in how socio-technical systems perform, about their prevailing performance patterns. These patterns encapsulate the heuristics and habits that people over time have found useful in their work.

Knowledge about systems

The term *system* comes from the Latin word *systēma*, which itself comes from the Greek σύστημα *(systēma)*, "whole concept made of several parts or members, system," or, literally, a "composition." The term is used to describe a set of entities that on the one hand can be distinguished individually and on the other are related to each other in such a way that it makes sense to consider them together rather than separately (cf. the definition by Hall and Fagen used in Chapter 5 or Deming's definition that a system is "a network of interdependent components that work together to try to accomplish the aim of the system" [Deming, 1994, p. 95]). A system will also require ways or modes of description that differ from what is applicable to or meaningful for any of the constituent entities.

The term *the system as a whole* is often used to emphasise that it is necessary as well as useful to adopt an inclusive or comprehensive rather than particularistic view. The drawback with this term is that a system is never whole in the sense that it is complete in itself without in some way being related to or dependent on something else. As discussed in Chapter 4 (thermodynamic entropy), there will, in practice, always be a controlling system, a controlling system beyond that, and so on; or a system may be part of another system, as in the concept of system of systems. Talking about the system as a whole also elegantly skirts the problem of defining the boundary. System as a whole suggests that there is no need to consider anything else – in particular that there is nothing "outside" the system, no surroundings or context. Yet a system will always have a boundary, and it is essential that this is defined and known. It certainly has consequences for what can be controlled and what cannot.

The nature of systems: structure and function

When a system commonly is defined as "a set of objects together," it makes sense to refer to a system's structure, to how the parts or elements are arranged or put together. This is evidently reasonable for systems that are composed of physical parts or entities regardless of whether the system is fixed (a house) or movable (an aeroplane), simple (a water clock) or complex (a power plant). But it does not really make sense to talk about the structure of social systems such as organisations even though we often loosely refer to their structure or architecture. Organisations are, of course, normally pictured as a hierarchical arrangement of divisions, departments, units, and sections where the connecting lines represent authority, communications, etc. (cf. Figure 1.1). The organisational structure is said to determine how roles and responsibilities are assigned, how they are controlled and coordinated, how information flows between the different parts of the organisation, and so on.

Yet rather than referring to an organisation's structure or architecture it is more useful to describe it in terms of what it does and hence in terms of the functions required for the intended overall performance and how they are organised. The origin of "to organise" is the Latin *organum*, which means "instrument" or "tool." It is clearly more important what a tool can *do* than what it *is*. As an alternative to structure, a system or an organisation can therefore be defined as the set of coupled functions needed to produce a certain performance. The functional definition is not only more interesting but also more useful. And it actually makes it possible to refer to an organisation's structure, although in a different sense.

The number of functions that can be found in an organisation is usually so large that it is sensible to think of it as sets of functions rather than as individual functions. At any given time some of these sets of functions will be active because they are part of carrying out a specific, ongoing activity. Within an active set the functions will be closely coupled and will therefore continuously change in relation to each other. Furthermore, there may be several such sets because several things may be happening at the same time in even a moderately large organisation. Relative to the specific activity these functions can be called the foreground functions or the set, or sets, of foreground functions. There will be other sets of functions – probably the majority – that provide the necessary support or background for the foreground functions, such as the supply lines of an army or the supply stream for just-in-time production. These functions are also active but usually more regular and therefore stable in the sense of being more predictable. There may be yet other functions that are needed in the longer term but that, relative to a given focus of activity, may be best characterised as dormant. March and Simon (1993) used this approach to describe an organisation and to point out that only a small number of functions at any point in time would be what they called adaptive, i.e., undergoing change; the remainder would, relatively speaking, be stable (the term they used was *givens*). They went on to argue:

> Organisation structure consists simply of those aspects of the pattern of behaviour in the organisation that are relatively stable and that change only slowly.
>
> (March & Simon, 1993, p. 191)

The basic argument is that a system or organisation cannot function if everything changes at the same time, i.e., if everything is in a flux. There must be some functions or sets of stable (background) functions that can serve as a basis of reference for change and adaptation, and these must be more stable than the active (foreground) functions that are undergoing change. The background functions still change, though at a slower rate, and are not coupled as tightly as the foreground functions. This way of looking at what goes on may also be used to describe an organisation in terms of multiple levels of functions, each level more stable or changing more slowly than the one preceding it. March and Simon suggested that "short-run adaptiveness corresponds to what we ordinarily call problem solving, long-run adaptiveness to learning" (Ibid., p. 192). In the terms of the voyage metaphor, the changes are as essential for the voyage itself as the changes to the current position. But the target must be stable, at least during major segments of the voyage, since there otherwise would be no way of knowing whether the changes in position went in the right direction.

By thinking about sets of functions with different degrees of stability it becomes possible to think about organisational structures without the need of decomposition and components. The structure no longer refers to how the parts or components are formally arranged, but rather to how the functions work together to achieve something. The needed stability is therefore relative to the duration of the activity (the definition of foreground and background functions is also relative to activity that is in focus). Some sets of functions need to be stable as long as the activity takes place as a whole, some only for shorter segments.

On boundaries

Calling something a system – or even a system of systems – means that it can be separated or distinguished from what is not the system. The latter is usually called the environment or (in this book) the surroundings. A system is always defined relative to its surroundings, which effectively means that there must be a boundary. The boundary makes it possible to refer to the system as that which is inside the boundary and the surroundings as that which is outside the boundary. The system is clearly finite. Whether the same goes for the surroundings is a different matter. They may in principle be as infinite as the universe, although there pragmatically will always be a limit. A boundary has traditionally been defined physically in terms of structures/components/parts and therefore thought to be relatively stable. This made obvious sense at the time when productivity, quality, and safety became major issues since the focus in all cases was some kind of work activity in well-structured environments. But that was a century ago, and today it makes more sense to describe a boundary functionally in terms of processes and their stability or variability. This makes the boundary more fluid and permeable. The latter is important because matter, energy, and especially information must be able to pass through the boundary; otherwise the system would be a closed system that could not be influenced by the surroundings and hence impossible to control.

> For a given system, the environment is the set of all objects a change in whose attributes affect [*sic*] the system and also those objects whose attributes are changed by the behaviour of the system.
>
> (Hall & Fagen, 1968, p. 83)

Fragmented change management schemes, such as PDSA, do not explicitly refer to a boundary. They are nevertheless defined indirectly via the fragmentation of scope these schemes imply. Planning a change always takes place relative to a limited scope and hence assumes a boundary. As described in Chapter 6, a temporarily defined boundary is a logical consequence of the *ceteris paribus* principle, the need to consider the context or surroundings as predictable and stable. It is impossible to make a change or to plan an improvement without adequate knowledge of the conditions or factors that play a role and that may have consequences for the outcome, either in achieving what was intended or in preventing it: "[d]ifferences in implementation may be due to differences in context" (Øvretveit, 2011). The context in the classical change paradigms is, however, only vaguely defined as "all factors that are not part of a quality improvement intervention itself" (Øvretveit, 2011, p. i18). Those parts of the surroundings that may influence the effectiveness of improvements, often called the conditions for improvement, are vaguely defined as "those internal to the implementing organisation (e.g., information technology) and those external to it (e.g., payment and regulation systems), and are made by, and operate on, different levels of the health system" (Ibid.).

It is nevertheless necessary not just to acknowledge that there is a boundary but also to be able to define it more clearly in order to distinguish between what is internal – the system inside the boundary – and what is external – the surroundings or context and, in principle, the rest of the world. In both cases it is also important that enough is known about what happens and how it happens in either place. It is obviously indispensable relative to what is inside the system since an understanding of the system dynamics or of what actually goes on is a prerequisite for suggesting specific interventions to bring about intended changes and for "changing the position." In common parlance this is referred to as the model of the system, something that was discussed in Chapters 4 and 6. But it is equally indispensable relative to the surroundings since what happens there constitutes an important source of the noise that may weaken, distort, or quell – but sometimes also amplify or strengthen – the signal.

A boundary will always be relative rather than absolute because it must be defined according to a set of criteria that depend on the purpose of the analysis and hence on the system's function rather than its structure. From a general point of view the distinction implies that the surroundings both provide the inputs to the system and react to outputs from the system. It may reasonably be asked why (some of) that which reacts to the system, as described by the above definition, is considered as the system's surroundings instead of simply being included in the system. Yet to automatically do so would only result in forever moving the boundary outwards without solving the demarcation problem. It may also rightfully be asked whether the surroundings themselves have a boundary or whether they are infinitely large. Answers to these questions can best be found by taking a pragmatic rather than formal stance. Indeed, the boundary of a system cannot be determined in an absolute sense but only relative to a specific purpose or aim. The determination is, in practice, made by limiting the system to those functions that are necessary and sufficient for its performance and excluding those that are not, all relative to the purpose of the analysis. (The problem with this is that the necessary set of functions is constantly growing due to the relentless enlargement view of work described in Chapter 6.)

The boundary of a system or organisation can be defined pragmatically by considering two aspects. The first is whether a function is important for the system, i.e., whether a function constitutes a significant source of variability for the system. The second is whether it is possible for the system to effectively control a function so that its variability remains within a pre-defined and acceptable range. By using these criteria, we may propose a pragmatic definition of how to determine the boundary of a system, as indicated in Table 7.1.

Of the four categories in Table 7.1, the third is in many ways the most important. If functions that are important for the ability to control a system, or to manage changes to a system, cannot be controlled they will constitute a source of noise that may make it impossible to achieve what was planned and intended. It is irresponsible to neglect or disregard these by simply assuming that the *ceteris paribus* principle is valid, that "everything else" is equal. Instead, every sensible effort should be made to bring them under control and hence ensure that they are within the system's boundary. If this is not possible, the only reasonable alternative is to revise the extent of the planned change and to adjust the ambitions accordingly even if it clashes with the "imperious immediacy of interest."

Table 7.1 A functional definition of *boundary*

	Functions that are important for the ability of the system to maintain control	Functions that are of no consequence for the ability of the system to maintain control
Functions that can be effectively controlled	1. Functions are included in the system.	2. Functions may be included in the system.
Functions that cannot be effectively controlled	3. Functions are not included in the system.	4. Functions are excluded from the description as a whole.

Knowledge about variability

Predictability is an essential condition for control and therefore also for effective management. The more unpredictable or irregular something is the longer it takes before a response can be made – and losing time can be critical for control. Conversely, when events and outcomes are predictable or regular, responses can follow quickly because they have been prepared (cf. Westrum, 2006). Similarly, if the outcome of an intervention is as expected, *pace* the confirmation bias, then change management can proceed in an orderly way. If not, the stepwise improvement cycle will be disrupted and may perhaps have to be abandoned. To avoid surprises it is therefore necessary to ensure predictability by avoiding variability – or so the traditional reasoning goes. The legacy of both quality and safety points to standardisation and compliance as the solution (cf. Chapter 2).

Assignable versus chance causes

Another legacy from quality and safety is the belief that all outcomes have their causes and that unwanted outcomes can therefore be avoided by finding their causes and eliminating them. This can be traced back at least to David Hume's rules (Chapter 6, especially rule #4) and can also be recognised in the present-day zero accident vision (e.g., Zwetsloot et al., 2013). The legacy of safety is simply the notion of the root cause and the associated analysis method. The legacy of quality is a little less straightforward – namely, the notion of assignable (or special) and chance (or common) causes.

As discussed in Chapter 2, Shewhart argued that there were two types of causes: assignable and chance. When quality exceeded the upper or lower control limits the explanation was to be found in assignable causes, which in practice meant that the benefit in finding (and dealing with) the causes would be larger than the costs. But this was a pragmatic rather than theoretical distinction, as the following passage makes abundantly clear:

> It shows that when points fall outside the limits, experience indicates that we can find assignable causes, but it does not indicate that when points fall within such limits, we cannot find causes of variability.
>
> (Shewhart, 1931, p. 19)

Deming made the same argument, except that he called chance causes common causes and assignable causes special causes.

> Common causes of variation produce points on a control chart that over a long period all fall inside the control limits. Common causes of variation stay the same day to day, lot to lot. A special cause of variation is something special, not part of the system of common causes.
>
> (Deming, 1994, p. 174)

Chance or common causes thus represented the "natural" variability of the processes or activities in focus, whereas assignable or special causes represented specific conditions that could be identified and brought under control. The difference was therefore not in the "nature" of the variability but in its amplitude and perhaps also frequency. (Chance and assignable causes could with equal justification be called acceptable and unacceptable, or even affordable and unaffordable, causes.) Change management approaches such as PDSA are based on Shewhart's hypothesis and therefore rely on the reality of assignable causes, the planned change itself being a particularly important one. The planned change will, as an intentionally assignable cause, lead to the intended effect because the system or conditions are deterministic. There may be chance variations (chance causes), but they are – by definition – so small, or rather their consequences are so small, that they can be disregarded in accordance with the *ceteris paribus* principle. The "check" in the PDSA cycle therefore serves to confirm the underlying hypothesis; namely, that there was no unknown assignable cause(s) at play.

By contrast, a contemporary perspective argues that variability is due to the performance adjustments that are the very basis for performance, for work that goes well as well as work that fails (Hollnagel, 2014). According to this perspective Shewhart's hypothesis is wrong and chance causes are not really chance or stochastic in the sense of being unpredictable. It is therefore necessary to revise the tradition of distinguishing between the two types of causes. As argued in Chapter 6 (the section "Signal and noise – again!"), both types of causes have the same basis or origin; namely, the adjustments that are made on all levels of an organisation and that are needed for work to go well. Chance causes consequently do not depend on chance and assignable causes are not special at all. On the contrary, it is only by coincidence that they become so large that they are classified as assignable. Therefore, rather than looking for assignable or special causes, corresponding to what in safety are called root causes, the focus should be on the conditions that make adjustments necessary and that, from time to time, through a process known as functional resonance (Hollnagel, 2012), may lead to adjustments so large that they are perceived as distinct causes. Indeed, the philosophy of eliminating assignable causes is misguided because it is based on the hypothesis of different causes – that unacceptable and acceptable outcomes happen for different reasons. From a Safety-II perspective they both happen for the same reasons; namely, the performance adjustments that are necessary for any kind of activity. The difference is in the magnitude of the outcomes, not the source. Eliminating assignable causes therefore erodes the basis for the performance adjustments that are needed for work to go well.

Another facet of this is the question of causality versus emergence. The basic idea of assignable causes assumes that outcomes are resultant rather than emergent in the sense that it is possible to trace the steps of the process and see the product as the result of specific factors or causes. This obviously made sense in the context of manufacturing and production since the systems in these cases were designed and built according to a plan. But the assumption does not hold for more dynamic and intractable systems. If we assume that some outcomes are emergent rather than resultant, or that they at least require a non-linear version of causality, then the variability in outcomes must be explained differently.

Variability as an asset

Resilience engineering and Safety-II maintain that performance adjustments (performance variability) are necessary for work to go well but also that they every now and then lead to unexpected and unwanted outcomes. Variability is a fundamental characteristic of socio-technical systems, and since no system can construct, operate, or maintain itself, all systems are socio-technical systems in one way or another. Performance adjustments are not only inevitable, due to the nature of the systems and to psychological fragmentation, they are also indispensable because plans and preparations will always be incomplete and imperfect. There will never be sufficient time and resources to get complete information about a situation. (The alternative solution is to reduce the variability of the situation so that the existing knowledge is sufficient – in other words, to force the real world to correspond to the world-as-imagined and ditto for work-as-imagined. However, experience shows that this solution does not work well in practice.) Because every situation is underspecified, the plans will never be precise. There will therefore always be unexpected situations or conditions during work, and to overcome these performance adjustments are required. The potential to adjust performance proactively, in anticipation, both improves the solution and exacerbates the problem. If our "guesses" are correct, then something is gained; if they are incorrect, then something will be lost. The solution to "play it safe" by only responding or adjusting when the situation has become sufficiently clear – which inevitably requires some kind of judgement – will eventually mean that we lag behind and hence have even more surprises and even less time to try to overcome or compensate for them.

The answer is, again, relatively simple, at least in principle. The variability and apparent non-determinism are by nature functional and purposeful rather than stochastic. They are purposeful given that they are limited by how well others are able themselves to control what they do and to predict or anticipate the consequences of their actions. Because they are not random, they are not completely non-deterministic. But the question remains how we can best understand and hence improve the ability to retain control. This is why knowledge about variability is an essential part of the nexus of necessary knowledge.

Knowledge about performance patterns

The third question, or third area of knowledge, is about the characteristics of human performance. This can be understood in two different ways: as referring to *why* people do what they do (also called the drivers or motivations of performance) or as referring to *how* people do what they do. The first is what puzzled Taylor when he found it hard to understand why the people he worked with did not produce more than one third of a good day's work. The same kind of puzzlement can also be found in later studies of loss aversion and "narrow bracketing," which describes how people make decisions based on perceived gains instead of perceived losses (Kahneman & Tversky, 1979).

In relation to synesis, the second version of the question is, however, more interesting. To manage an organisation, to manage a change, and, in particular, to be able to predict what will happen in response to an intervention, it is necessary to know the characteristics or patterns of human performance and to understand how they arise. People are emphatically not machines or robots that can be instructed or programmed to do as required, which is why the distinction between Work-as-Imagined and Work-as-Done is so important. People do what makes sense to them rather than what makes sense to those who manage them. It is essential to acknowledge that and work with it rather than against it or deny it or try to overrule it.

The necessity of adjustments

The fragmentation of issues that is a problem for management can also be a problem for the individual. Human performance must never satisfy just one criterion even if an organisation's manifesto or espoused values pretend it is so. What people do always has to meet multiple, changing, and often conflicting criteria of performance. Some of these come from the organisation as such; some from the social settings of the work, group expectations, and the like; and some from individual ambitions and expectations of work – the shared assumptions and levels of aspiration or *anspruchsniveau*. The various versions of organisational culture have tried to capture this by defining culture as "the way we do things here."

Humans usually cope with the multiple demands by adjusting what they do and how they do it to match the current conditions. This has been described in several ways by terms such as *adaptation, optimisation, satisficing, suffisance, minimising cognitive effort*, and *minimising workload* and analysed in terms of various forms of trade-offs (Hoffman & Woods, 2011). As a starting point, it is reasonable to assume that people constantly try to achieve an acceptable balance or trade-off between efficiency and thoroughness – between resources and demands, where both may vary over time (Hollnagel, 2009). On the one hand, they genuinely try to do what they are supposed to do – or at least what they believe it is reasonable to do – and to be as thorough as the situation allows. On the other hand, they try to do this as efficiently as possible, which means avoiding any unnecessary effort and waste of time or resources.

In trying to achieve this trade-off, people rely on the relative stability of whatever happens around them, on the "aspects of the pattern of behaviour in the organisation that are relatively stable and that change only slowly" (March & Simon, 1993, p. 191). If the work situation was totally unpredictable it would be impossible to take any shortcuts or, indeed, even to learn how work might be done in a more efficient manner. Conversely, when the work environment has some measure of regularity or stability – which it always has – it becomes predictable. This regularity is in turn due to the persistent patterns of performance and the trade-offs people use to optimise performance, thereby closing the loop. Human performance is on the whole systematic, but it is heuristic rather than algorithmic. (A heuristic is any practical method that is not guaranteed to be optimal, perfect, or rational but that nevertheless is sufficient to reach an immediate, short-term goal.) The psychological reasons for fragmentation, or the simple facts that govern our thinking, mean that these heuristics are found in work throughout an organisation, including change management.

Regularity and efficiency

Human performance is efficient because people quickly learn to disregard those aspects or conditions that, under normal circumstances, are insignificant. This adjustment is not only a convenient ploy for the individual but also a necessary condition for the organisation as a whole. Just as individuals adjust their performance to avoid unnecessary efforts, so, too, does an organisation. This creates a functional reciprocity that is essential for understanding how things happen and therefore how they can be managed and changed. The performance adjustment on an organisational or system level cannot be effective unless the aggregated effects of individual performance are relatively stable. By contrast, the stable and efficient performance of the organisation provides the regularity of the work environment that is a precondition for the performance adjustments individuals make and hence for the effectiveness of everyday work.

As far as individual performance is concerned, adjustments are the norm rather than the exception. Indeed, typical, or "normal," performance is not that which is prescribed by rules and regulation, but rather that which takes place as a result of the adjustments, i.e., the equilibrium that reflects the regularity of the work environment. Regularity is thus a precondition for efficiency, just as efficiency is a precondition for regularity. Understanding that is an essential contribution to the nexus of necessary knowledge.

Final words

We should be concerned about any issue that is essential for the existence and long-term sustainability of an organisation – but not in isolation and not in a fragmented way. On the contrary, it is absolutely necessary to overcome the consequences of fragmentation, whether it has historical or psychological reasons. To do so requires changing the ways that common problems and issues are addressed, first and foremost by being realistic about the scope and time window of changes. We should be concerned about productivity and do whatever we can to ensure both short- and long-term productivity goals – but not in isolation and not in a fragmented way. We should be concerned about quality and do our best to ensure that we achieve the quality we need – but not in isolation and not in a fragmented way. We should be concerned about safety when things go well and when they fail – but not in isolation and not in a fragmented way. And, finally, we should be concerned about reliability and try to ensure the necessary reliability of functions in every aspect of system performance – but not in isolation and not in a fragmented way.

> The real trouble with this world of ours is not that it is an unreasonable world, nor even that it is a reasonable one. The commonest kind of trouble is that it is nearly reasonable, but not quite. Life is not an illogicality; yet it is a trap for logicians. It looks just a little more mathematical and regular than it is; its exactitude is obvious, but its inexactitude is hidden; its wildness lies in wait.
>
> (G. K. Chesterton, 2008; org. 1909)

References

Adamski, A. & Westrum, R. (2003). Requisite imagination: The fine art of anticipating what might go wrong. In E. Hollnagel (Ed.), *Handbook of cognitive task design* (pp. 193–220). Mahwah, NJ: Lawrence Erlbaum Associates.

Advisory Group on Reliability of Electronic Equipment (AGREE). (1957). *Reliability of military electronic equipment.* U.S. Department of Defense. U.S. Government Printing Office, Washington, DC.

Aristotle. (1970). *Aristotle's physics: Books 1 & 2.* Oxford: Clarendon Press.

Ashby, W. R. (1957). *An introduction to cybernetics.* London: Chapman & Hall, Ltd.

Beer, S. (1984). The viable system model: Its provenance, development, methodology and pathology. *Journal of the Operational Research Society, 35*(1), 7–25.

Berengueres, J. (2007). *The Toyota production system re-contextualized.* www.lulu.com.

Bird, F. E. (1974). *Management guide to loss control.* Atlanta, GA: Institute Press.

Boyd, J. (1987). *Destruction and creation.* US Army Command and General Staff College. http://goalsys.com/books/documents/DESTRUCTION_AND_CREATION.pdf (Accessed June 1, 2020).

Brehmer, B. (2005). The dynamic OODA loop: Amalgamating Boyd's OODA loop and the cybernetic approach to command and control. In *Proceedings of the 10th international command and control research technology symposium*, 13–16 June 2005, MacLean, VA, 365–368.

Carroll, J. M. & Campbell, R. L. (1988). *Artifacts as psychological theories: The case of human: Computer interaction.* IBM Research Report RC 13454, Watson Research Center, Yorktown Heights, New York.

Chesterton, G. K. (2008). *Orthodoxy.* West Valley City, UT: Waking Lion Press.

Conant, R. C. & Ashby, W. R. (1970). Every good regulator of a system must be a model of that system. *International Journal of Systems Science, 1*(2), 89–97.

Coombs, C. H., Dawes, R. M., & Tversky, A. (1970). *Mathematical psychology.* Englewood Cliffs, NJ: Prentice Hall, Inc.

Dekker, S. W. & Woods, D. D. (2002). MABA-MABA or abracadabra? Progress on human: Automation co-ordination. *Cognition, Technology & Work, 4*(4), 240–244.

Deming, W. E. (1994). *The new economics for industry, government, education* (2nd ed.). Cambridge, MA: The MIT Press.

Donnelly, P. & Kirk, P. (2015). Use the PDSA model for effective change management. *Education for Primary Care, 26*(4), 279–281.

Drury, H. B. (1918). Scientific management: A history and criticism. *Studies in History, Economics, and Public Law, 65*(2), Whole Number 157.

Emery, F. E. & Trist, E. L. (1965). The causal texture of environments. *Human Relations, 18*, 21–32.

Fitts, P. M. (1951). *Human engineering for an effective air-navigation and traffic-control system.* Washington, DC: National Research Council.

Forrester, J. W. (1971). Counterintuitive behavior of social systems. *Technological Forecasting and Social Change, 3*, 1–22.

Gribbin, J. (2003). *The scientists: A history of science told through the lives of its greatest inventors.* London: Penguin Books, Ltd.

Hall, A. D. & Fagen, R. E. (1968). Definition of system. In W. Buckley (Ed.), *Modern systems research for the behavioural scientist.* Chicago: Aldine Publishing Company.

Harvey, L. & Green, D. (1993). Defining quality. *Assessment & Evaluation in Higher Education, 18*(1), 9–34.

Heinrich, H. W. (1928). The origin of accents. *The Travelers Standard, 16*(6), 121–137.

Heinrich, H. W. (1929). The foundation of a major injury. *The Travelers Standard, 17*(1), 1–10.

Heinrich, H. W. (1931). *Industrial accident prevention: A scientific approach.* New York: McGraw-Hill.

Hoffman, R. R. & Woods, D. D. (2011). Beyond Simon's slice: Five fundamental trade-offs that bound the performance of macrocognitive work systems. *IEEE Intelligent Systems, 26*(6), 67–71.

Hollnagel, E. (2003). Prolegomenon to cognitive task design. In E. Hollnagel (Ed.), *Handbook of cognitive task design* (pp. 3–16). Mahwah, NJ, USA: Lawrence Erlbaum Associates.

Hollnagel, E. (2009). *The ETTO principle: Efficiency-thoroughness trade-off: Why things that go right sometimes go wrong.* Aldershot, UK: Ashgate.

Hollnagel, E. (2012). *FRAM: The functional resonance analysis method: Modelling complex socio-technical systems.* Farnham, UK: Ashgate.

Hollnagel, E. (2014). *Safety-I and Safety-II: The past and future of safety management.* Farnham, UK: Ashgate.

Hollnagel, E., Pariès, J., Woods, D. D., & Wreathall, J. (Eds.). (2011). *Resilience engineering perspectives volume 3: Resilience engineering in practice.* Farnham, UK: Ashgate.

Hollnagel, E. & Woods, D. D. (2005). *Joint cognitive systems: Foundations of cognitive systems engineering.* Boca Raton, FL: CRC Press/Taylor & Francis.

Hollnagel, E., Woods, D. D., & Leveson, N. G. (2006). *Resilience engineering: Concepts and precepts.* Aldershot, UK: Ashgate Publishing Ltd.

Hume, D. (1985). *A treatise of human nature.* London: Penguin Books.

Institute for Healthcare Improvement. (2019). *How to improve.* www.ihi.org/resources/Pages/HowtoImprove/ScienceofImprovementTestingChanges.aspx (Accessed November 11, 2019).

International Transport Forum. (2018). *Safety management systems: Summary and conclusions.* Roundtable 172. Paris: OECD.

ISO. (2015). *Quality management systems: Requirements* (ISO 9001:2015(en)). Geneva, Schweitz: International Organization for Standardization. www.iso.org/obp/ui/#iso:std:iso:9001:ed-5:v1:en (Accessed January 8, 2020).

James, W. (1890). *The principles of psychology.* London: Macmillan and Co.

Kahneman, D. & Tversky, A. (1979). Prospect theory: An analysis of decision under risk. *Econometrica, 47*(2), 263–291.

Klein, G. A., Orasanu, J. M., Calderwood, R., & Zsambok, C. (1993). *Decision making in action: Models and methods.* Norwood, NJ: Ablex Publishing Corporation.

Kurtz, C. & Snowden, D. J. (2003). The new dynamics of strategy: Sense-making in a complex and complicated world. *IBM Systems Journal, 42*(3), 462–483.

Leveson, N. G. (1992). High-pressure steam engines and computer software. In *Proceedings of the 14th international conference on software engineering* (pp. 2–14). New York, NY: Association for Computing Machinery.

Lewes, G. H. (1875). *Problems of life and mind.* Boston: James R. Osgood and Company.

Lewin, K. (1946). Action research and minority problems. In G. W. Lewin (Ed.), *Resolving social conflict.* London: Harper & Row.

Lewin, K. (1951). Frontiers in group dynamics. In D. Cartwright (Ed.), *Field theory in social science.* New York: Harper & Row.

Lindblom, C. E. (1959). The science of "muddling through". *Public Administration Review, 19*(2), 79–88.

Lofquist, E. A. & Lines, R. (2017). Keeping promises: A process study of escalating commitment leading to organizational change collapse. *The Journal of Applied Behavioral Science, 53*(4), 417–445.

March, J. G. & Simon, H. A. (1993). *Organizations* (2nd ed.). Cambridge, MA: Blackwell Business.

Maruyama, M. (1963). The second cybernetics: Deviation-amplifying mutual causal processes. *American Scientist, 5*(2), 164–179.

Maurino, D. (2017). *Why SMS: An introduction and overview of safety management systems (SMS).* Draft Discussion Paper for the Roundtable on Safety Management Systems, March 23–24. Paris: OECD, International Transportation Forum.

Merton, R. K. (1936). The unanticipated consequences of purposive social action. *American Sociological Review*, *1*(6), 894–904.

Milgram, S. (1967). The small world problem. *Psychology Today*, *2*(1), 60–67.

Miller, G. A. (1956). The magical number seven, plus or minus two: Some limits on our capacity for processing information. *Psychological Review*, *63*(2), 81–97.

Miller, J. G. (1960). Information input overload and psychopathology. *American Journal of Psychiatry*, *116*(8), 695–704.

Miller, J. G. (1978). *Living systems*. New York: McGraw-Hill.

Moen, R. & Norman, C. (2009). *Evolution of the PDCA cycle*. Paper delivered to the Asian Network for Quality Conference in Tokyo, Japan, on September 17.

Nietzsche, F. (2007; org. 1895). *Twilight of the idols*. Published by Wordsworth Editions, UK.

Øvretveit, J. (2011). Understanding the conditions for improvement: Research to discover which context influences affect improvement success. *BMJ Quality & Safety*, *20*(Suppl. 1), i18ei23. doi:10.1136/bmjqs. 2010.045955.

Paxton, L. J. (2007). "Faster, better, and cheaper" at NASA: Lessons learned in managing and accepting risk. *Acta Astronautica*, *61*(10), 954–963.

Perrow, C. (1984). *Normal accidents: Living with high risk technologies*. New York: Basic Books, Inc.

Pigeau, R. & McCann, C. (2002). Re-conceptualizing command and control. *Canadian Military Journal*, *3*(1), 53–63.

Rasmussen, J. (1986). *Information processing and human-machine interaction*. New York: North-Holland.

Reason, J. T. (1988). Cognitive aids in process environments: Prostheses or tools? In E. Hollnagel, G. Mancini, & D. D. Woods (Eds.), *Cognitive engineering in complex dynamic worlds*. London: Academic Press.

Reason, J. T. (2000). Safety paradoxes and safety culture. *Injury Control & Safety Promotion*, *7*(1), 3–14.

Roberts, K. H. (1989). New challenges in organizational research: High reliability organizations. *Organization & Environment*, *3*(2), 111–125.

Roberts, K. H. (1990). Some characteristics of one type of high reliability organization. *Organization Science*, *1*(2), 160–176.

Saleh, J. H. & Marais, K. (2006). Highlights from the early (and pre-) history of reliability engineering. *Reliability Engineering and System Safety*, *91*, 249–256.

Schützenberger, M. P. (1954). A tentative classification of goal-seeking behaviours. *Journal of Mental Science*, *100*, 97–102.

Shewhart, W. A. (1931). *The economic control of quality of manufactured product*. New York, NY: D. Van Nostrand Company.

Shewhart, W. A. (1939). *Statistical method from the viewpoint of quality control*. Department of Agriculture. Washington, DC: Graduate School of the Department of Agriculture.

Simon, H. A. (1956). Rational Choice and the Structure of the Environment. *Psychological Review*, *63*(2), 129–138.

Smith, A. (1986). *The wealth of nations*. Harmondsworth, UK: Penguin Classics.

Stevenson, R. L. (1969). *Treasure Island*. Harmondsworth, UK: Penguin Books Ltd.

Swain, A. D. & Guttmann, H. E. (1983). *Handbook of human reliability analysis with emphasis on nuclear plant applications*. NUREG/CR-1278. Albuquerque, NM: Sandia Laboratories.

Tajfel, H. & Turner, J. C. (1986). The social identity theory of intergroup behaviour. In S. Worchel & W. G. Austin (Eds.), *Psychology of intergroup relations* (pp. 7–24). Chicago, IL: Nelson-Hall.

Taylor, F. W. (1911). *The principles of scientific management*. New York, NY: Harper & Brothers.

Tolstoy, L. (1993). *War and peace*. Ware, Hertfordshire: Wordsworth Classics.

Tversky, A. & Kahneman, D. (1974). Judgment under uncertainty: Heuristics and biases. *Science*, *185*(4157), 1124–1131.

von Clausewitz, C. (1989). *On war*. Princeton, NJ: Princeton University Press.

Weick, K. E. & Sutcliffe, K. M. (2001). *Managing the unexpected: Resilient performance in an age of uncertainty*. San Francisco, CA: Jossey-Bass.

Westrum, R. (2006). A typology of resilience situations. In E. Hollnagel, D. D. Woods, & N. Leveson (Eds.), *Resilience engineering: Concepts and precepts* (pp. 55–65). Aldershot, UK: Ashgate.

Wikipedia. *Quality management system.* https://en.wikipedia.org/wiki/Quality_management_system.

Woods, D. D. & Watts, J. C. (1997). How not to have to navigate through too many displays. In *Handbook of human-computer interaction* (pp. 617–650). Amsterdam, the Netherlands: Elsevier Science B.V.

Wright, R. (2004). *A short history of progress.* Toronto: Anansi Press.

Zwetsloot, G. I. J. M., Aaltonen, M., Wybo, J.-L., Saari, J., Kines, P., & Op De Beeck, R. (2013). The case for research into the zero accident vision. *Safety Science, 58,* 41–48.

Index

Printed in the United States
by Baker & Taylor Publisher Services